Animal Models in
Human Psychobiology

Animal Models in Human Psychobiology

Edited by

George Serban
Kittay Scientific Foundation

and

Arthur Kling
Rutgers Medical School

With a Foreword by
José M. R. Delgado

PLENUM PRESS · NEW YORK AND LONDON

Library of Congress Cataloging in Publication Data

Main entry under title:

Animal models in human psychobiology.

Proceedings of the Second International Symposium of the Kittay Scientific
Foundation, held Mar. 24-26, 1974 in New York City.
Includes bibliographies and index.
1. Psychobiology, Experimental–Congresses. 2. Laboratory animals–Congresses.
3. Psychology, Comparative–Congresses. I. Serban, George, 1926-   II. Kling,
Arthur, 1929-   III. Kittay Scientific Foundation. [DNLM: 1. Animals, Lab-
oratory–Congresses. 2. Behavior–Congresses. 3. Neurochemistry–Congresses. 4.
Psychophysiology–Congresses. 5. Neurophysiology–Congresses. WL102 A598
1974]
QP356.A73   616.8'9'027   75-40449
ISBN 0-306-30864-9

Proceedings of the Second International Symposium of the Kittay Scientific
Foundation on "Relevance of the Animal Psychopathological Model to the Human,"
held March 24-26, 1974 in New York, New York

We gratefully acknowledge the participation of Hoffman–La Roche and Dow
Chemical Companies and their contribution toward this symposium

Manuscripts prepared by Arlyne Zimmermann, Director of Communications,
Kittay Scientific Foundation

© 1976 Plenum Press, New York
A Division of Plenum Publishing Corporation
227 West 17th Street, New York, N.Y. 10011

United Kingdom edition published by Plenum Press, London
A Division of Plenum Publishing Company, Ltd.
Davis House (4th Floor), 8 Scrubs Lane, Harlesden, London, NW10 6SE, England

Foreword

In March, 1974, an International Symposium was held at the Harmonie Club in New York to discuss a highly pertinent problem in today's research: the "Relevance of the Animal Psychopathological Model to the Human." This meeting was sponsored by the Kittay Foundation, which brought together an outstanding group of scientists involved in widely different fields of research. This volume, it is hoped, will convey the tone of lively and cordial exchange between internationally renowned investigators, including Dr. I. Eibl-Eibesfeldt from Germany, Dr. Robert A. Hinde from England, Dr. Edward F. Domino from Michigan, and Dr. Pierre Pichot from France, Chairman of the Steering Committee.

In his welcoming address, Mr. Sol Kittay reminded us that man has achieved remarkable control over his environment but not over himself, and he suggested that we should reexamine our ancestral origins, and search in animal behavior for clues to the understanding of normal and abnormal behavior in man.

The contrast between the fast pace of industrial and technological revolutions and the slow advances in the improvement of human behavior is creating a growing danger. We are facing a world in which vast power is at the disposal of brains which have not yet learned to solve their social and ideological conflicts intelligently. Some learned and responsible groups, such as the Club of Rome, have voiced their alarm about the catastrophic course of mankind, plagued by overpopulation, shortages of food, energy, and raw materials, and threatened with an atomic "balance of terror."

Fortunately, there is growing awareness of the lack of balance between the physical and neurobehavioral sciences, recognizing that an effort of unusual proportions is needed to promote psychophysiological research and to find solutions to the new problems of rapidly evolving mankind. Appropriate technology is becoming available, and strategies must be devised to mobilize public opinion, talent, and resources. The present meeting was a step in that direction.

In psychophysiology, the use of biological or even mechanical models permits the control of variables, simplification of problems, repetition of experiments, and design of studies without the obvious constraints of research with human subjects. Fortunately, some physiological mechanisms are similar in widely differing species and, for example, much of our knowledge of nerve physiology, based on studies on frogs, squids, and lobsters, is fully applicable to mammals, including man.

Efforts are now being extended to find models for more complex behavior, and as reported in this symposium, the study of hyperkinetic dogs is relevant for the understanding and, ideally, for the therapy of hyperkinetic children, while baby monkeys deprived of their mothers may be suitable models for the problem of human babies suffering from similar maternal deprivation.

Some speakers expressed concern about the application of findings in lower animals to the higher ones, especially to man. For example, Dr. Hinde indicated that the results of some of his studies of mother–child separations were dissimilar in monkeys and man, while studies of the "nuclear family" in monkeys by Dr. Suomi and data on mother–baby relations presented by Dr. Bowlby were more relevant to human behavior.

In experimental design, the choice of animal species is of decisive importance. The rat is a very useful animal, as it is inexpensive, easy to handle and breed, and possesses an interesting behavioral repertoire. For these reasons, rats have been favored in psychological research to such an extent that, as Koestler has warned, we risk acquiring a ratomorphic concept of man. The preponderance of research in rats for budgetary reasons may not be fully justified, considering that the investment in equipment and personnel often far outweighs the cost and care of animals. The rat remains, however, a highly suitable subject for many experiments. In this symposium, for example, Dr. Weiss reported that a transitory stress-induced, neurochemical change in rats is responsible for the failure to cope adequately with a difficult stress situation. The demonstration that this mechanism is physiological and not psychological is most significant.

A growing research area in need of new animal models is the study of the neurochemical bases of behavior. Cholinergic, adrenergic, serotonergic, and other chemical systems are related to sleep–wakefulness cycles, to offensive–defensive manifestations, and perhaps to schizophrenia. As discussed at this meeting, neurochemical changes are also involved in stress and depression. In order to analyze chemical changes, it is usually necessary to kill the experimental animal, extract the brain, and homogenize the tissue. Results are therefore difficult to localize neuroanatomically and cannot be repeated on the same subject, while controls require an additional number of animals. In part, these problems may be solved by permanent implantation of push–pull cannulae in the brain, and this method is now used in several laboratories. In monkeys, for example, simultaneous perfusion of synthetic spinal fluid in several cerebral areas permits investigation of neurochemical cross-correlations related to different behavioral states in awake subjects performing psychological tests.

Another animal model with great potential in the study of the neurochemistry of behavior is the unanesthetized goat equipped with intracerebral chemitrodes. This animal has a "split brain circulation" due to its undeveloped vertebral arteries and functional circulatory separation between left and right cerebral hemispheres. It is therefore possible to compare simultaneously the neurochemical activity of each independent half of the brain, using one side as control.

A recurrent criticism in the use of laboratory animals is that the restraint and artificiality of the cage situation distort behavior and physiological data. It is maintained that experiments in confined subjects should not be compared with results obtained in free animals. This controversy may be clarified with with two types of experiments: (a) the use of wireless communication in completely free animals for both telemetry of physiological data and brain radio stimulation, and (b) the alternate study of the same animals first in the laboratory and then in free-ranging situations.

As emphasized by Dr. George Serban, Medical Director of the Kittay Foundation, psychology and psychiatry must attain greater depth in their scientific knowledge, and as the human brain is not an experimental subject, it is necessary to explore other primates extensively.

Brain research is at present a very active field—challenging and controversial— and the Kittay Foundation has played an important role in bringing together scientists of distinction—those already well established and some newly arrived— for the needed cross-fertilization and passing of the torch, stimulating future research and public awareness.

In the next few years, investigations of normal and abnormal function of the nervous system should provide new therapies for patients suffering from neurological disorders, and it may be hoped that a clearer understanding of human behavior, both individual and social, will give us solutions for the present conflicts of mankind, the bases of which are not only material, but mental, for the power center within man is his brain, creator of conflicts and solutions, of dreams and realities.

Madrid, Spain   José M. R. Delgado

Participants

Abdul H. Abballah
Research Specialist, Dow Chemical Company

Mary D. Salter Ainsworth
Professor, Department of Psychology, University of Virginia

John Bowlby
Tavistock Institute of Human Relations, School of Family Psychiatry & Community Mental Health

Francis Braceland
Editor, American Journal of Psychiatry; Senior Consultant, The Institute of Living

Eugene B. Brody
Professor & Chairman, Department of Psychiatry, University of Maryland

Richard Allan Chase
Associate Professor of Psychiatry & Behavioral Sciences, The Johns Hopkins University School of Medicine

Leonard Cook
Assistant Director of Pharmacology, Hoffman—La Roche

Samuel A. Corson
Professor of Psychiatry (Physiology) & Biophysics, The Ohio State University; Cerebrovisceral Physiology Laboratory

Borje Cronholm
Director, Department of Psychiatry, Karolinska Institute, Stockholm, Sweden

José M.R. Delgado
Professor of Physiology, Universidad Autonome Facultad de Medicine, Madrid, Spain

Edward F. Domino
Professor, Department of Pharmacology, University of Michigan Medical School

Irenäus Eibl-Eibesfeldt
Arbeitsgruppe für Humanethologie, Max Planck Institut für Verhaltensphysiologie, Federal Republic of West Germany

Frank R. Ervin
Professor of Psychiatry, Center for Health Sciences, University of California at Los Angeles

Alfred F. Freedman
Past President, American Psychiatric Association; Professor & Chairman, Department of Psychiatry, New York Medical College

Arnold Friedhoff
Professor & Director, Millhauser Laboratories; New York University Medical Center, School of Medicine

Bernard C. Glueck
Director of Research, The Institute of Living

Murray Glusman
Professor of Clinical Psychiatry, and Chief, Department of Behavioral Physiology, New York State Psychiatric Institute, Columbia University

Harry F. Harlow
Primate Laboratory & Department of Psychology, University of Wisconsin

Morris Herman
Acting Chairman, Department of Psychiatry, New York University Medical Center

Robert A. Hinde
MRC Unit on the Development & Integration of Behavior, University Sub-Department of Animal Behavior, Cambridge, England

William Holz
Senior Investigator, Smith, Kline & French Laboratories

Howard F. Hunt
Chief of Psychiatric Research (Psychology), New York State Psychiatric Institute, Columbia University

Samuel Irwin
Professor of Pharmacology, University of Oregon Medical School

Martin M. Katz
Chief, Clinical Research Branch, American College of Neuropsychology, National Institute of Mental Health

Arthur Kling
Professor, Department of Psychiatry, Rutgers Medical School

Lawrence C. Kolb
New York State Commissioner of Mental Hygiene; Professor of Psychiatry, College of Physicians & Surgeons, Columbia University

Seymour Levine
Professor & Director, Laboratory of Developmental Psychobiology, Stanford University

Franklin N. Marshall
Director of Pharmacology, Dow Chemical Company

Jules H. Masserman
Professor Emeritus of Psychiatry & Neurology, Northwestern University Medical School

Herbert Y. Meltzer
Associate Professor, Department of Psychiatry, University of Chicago

John Money
Professor of Medical Psychology, and Associate Professor of Pediatrics, The Johns Hopkins University School of Medicine

Dennis Murphy
Chief, Section on Neuropharmacology, Laboratory of Clinical Science, National Institute of Mental Health

Ronald Myers
Chief, Laboratory of Perinatal Physiology, National Institute of Neurological Diseases & Stroke

Pierre Pichot
Professor & Chairman, Department of Psychiatry, University of Paris

Morton F. Reiser
Professor & Chairman, Department of Psychiatry, Yale University Medical School

Harvey Shein
Associate Professor, Harvard Medical School; Clinical Director, McLean Hospital

Charles Stroebel
Director of Psychophysiological Laboratories, The Institute of Living

Stephen J. Suomi
Research Associate & Lecturer, Primate Laboratory & Department of Psychology, University of Wisconsin

S. S. Tennen
Senior Research Investigator, Searle Laboratories

Hugh A. Tilson
Senior Research Scientist, Bristol Laboratories

Louis Jolyon West
Professor & Chairman, University of California Center for Health Sciences

Jay M. Weiss
Associate Professor, The Rockefeller University

Contents

New Perspectives in Psychiatry: Relevance of the Psychopathological
Animal Model to the Human   1
George Serban, Pierre Pichot, Alfred F. Freedman, and Sol Kittay

I. INSTINCTUAL AND ENVIRONMENTAL LEARNING

Factors Affecting Responses to Social Separation in Rhesus Monkeys   9
Stephen J. Suomi

Human Personality Development in an Ethological Light. 27
John Bowlby

Discussion of Suomi and Bowlby Chapters 37
Mary D. Salter Ainsworth

Prenatal and Postnatal Factors in Gender Identity · 49
John Money

Workshop I (Moderated by Harry F. Harlow) 61
Harry F. Harlow and Stephen Suomi (Editors)

II. CONFLICT OF ADAPTATION TO CHANGED OR INDUCED ENVIRONMENTAL CONDITIONS

Phylogenetic and Cultural Adaptation in Human Behavior. 77
Irenäus Eibl-Eibesfeldt

Unpredictability in the Etiology of Behavioral Deviations 99
Jules H. Masserman

Animal Models of Violence and Hyperkinesis: Interaction of Psychophar-macologic and Psychosocial Therapy in Behavior Modification 111
Samuel S. Corson, E. O'Leary Corson, L. Eugene Arnold, and
Walter Knopp

Coping Behavior and Neurochemical Changes in Rats: An Alternative Explanation for the Original "Learned Helplessness" Experiments 141
Jay Weiss, H. I. Glazer, and L. A. Pohorecky

Discussion 175
Morton F. Reiser

Workshop II (Moderated by Howard F. Hunt) 181
Howard F. Hunt (Editor)

III. NEUROPHYSIOLOGICAL EXPERIMENTAL MODIFICATION OF THE ANIMAL MODEL AS APPLIED TO MAN

The Use of Differences and Similarities in Comparative Psychopathology .. 187
R. A. Hinde

Animal Models for Brain Research 203
José M. R. Delgado

Drug Effects on Foot-Shock-Induced Agitation in Mice 219
Samuel Irwin, Roberta G. Kinohi, and Elaine M. Carlson

Indole Hallucinogens as Animal Models of Schizophrenia 239
Edward F. Domino

Discussion 261
Arthur Kling

Animal Models for Human Psychopathology: Observations from the Vantage Point of Clinical Psychopharmacology 265
Dennis L. Murphy

Workshop III (Moderated by Ronald D. Myers) 273
Dennis Murphy (Editor)

Concluding Remarks 275
Borje Cronholm

The Significance of Ethology for Psychiatry 279
G. Serban

Index of Names 291

Subject Index 293

New Perspectives in Psychiatry: Relevance of the Psychopathological Animal Model to the Human

GEORGE SERBAN, PIERRE PICHOT, ALFRED F. FREEDMAN, AND SOL KITTAY

Biological and social behaviorists have been attempting to bridge the gap between natural and social sciences to achieve a better understanding of human behavior. It is refreshing to see the return of psychology to the natural sciences, based on a re-evaluation of basic concepts in man's behavior in relation to that of the animal, which will result in a more scientific orientation of the behavioral sciences. Unfortunately, psychiatry has placed itself in a pseudoscientific position as a result of its confusing and ambiguous theoretical tenets. Indeed, psychiatrists are a special breed of healers prepared by a medical model for a neurophysiological understanding of man, while they are faced, in the process of healing, with a complex psychosocial reality of man interacting with this environment. Most clinical schools of thought that attempted to create a psychological model of man based on introspection concluded with highly speculative theories, scientifically impossible to test. Conversely, experimental researchers, working in laboratories with animal models, attempt to make quali-

GEORGE SERBAN ● Medical Director, Kittay Scientific Foundation; PIERRE PICHOT ● Dept. of Psychiatry, University of Paris, Paris, France; ALFRED F. FREEDMAN ● Dept. of Psychiatry, N.Y. Medical College; SOL KITTAY ● President, Kittay Scientific Foundation.

1

fications on a human system of behavior based on mechanical concepts totally unsuited to man's psychosocial existence.

Between scientific fragments of information and metaphysical beliefs, psychiatry is still trying to find its way in understanding man and his irrationality. The seriousness of the problem is indicated in our difficulty in even defining the meaning of normalcy. Moreover, the gap between the scientific and mystical understanding of man is widening, due to the present tendency in psychiatry to attribute to man undefinable cosmic structures of energy, claimed to be reached through ESP. This is, in my opinion, the final admission of our ignorance in understanding the nature of man.

The ethologist, on the other hand, brings a fresh, new appreciation of animal behavior, which can indeed be useful for interpreting the human behavioral mechanism. In this context, Tinbergen gave us a good piece of advice by saying that it is not necessary to follow the ethologists, but to attempt to use their method instead.

In this respect, the application of the methodological laws used by ethologists to human behavior might eliminate metaphysical explanation so amply used in psychiatry as a cover-up for our ignorance. Perhaps in the future we will be able to meet the goal proposed approximately 50 years ago by Pavlov: "Only science, exact science about human nature itself and the most sincere approach to it by the aid of the more omnipotent scientific methods, will deliver man from his present gloom, and purge him from his contemporary shame in the sphere of interhuman relationships."

Behaviorism failed to give us this desired exact science of man—it resulted instead in a manipulative approach to the human organism, which was conceived of as a machine. Mechanistic psychology lost its support following the latest developments in physics, from which it drew its initial force.

The principle of uncertainty, the irreducible imprecision of physical measurement, along with Bohr's law of complementarity, changed our understanding of an exact mechanical science of man. This resulted in a reconciliation of mechanistic causality with purposive rationality applied to our behavioral science and served to reconcile the deterministic thesis of human nature with the cultural one.

Yet the environmentalists are still objecting strongly to any strict biological and instinctual conceptualization of human behavior based on millions of years of cultural evolution of mankind. The ethologists attempted to resolve the problem by trying to show the interaction between instinctual programming and social evolution, which is preprogrammed itself. This means that the process of man's evolution should be understood in the light of his adaptability to the environment, due to his complex ability to learn, memorize, and solve problems by abstraction and symbolization within his biological framework. This preprogrammed framework only permits diversities of approach within the same structural context.

In this sense, environmental evolution is determined by the biological basis of behavior, which in turn is influenced by environment. This complex inter-action between heredity and environment is sometimes almost impossible to separate, particularly since the environmental influences leave their imprints most significantly in man's early years of life.

Sir Charles Sherrington, in his Gifford Lecture of 1937–37 entitled "Man on His Nature," made the following statement, which takes on a new urgency for the development of a science of man. "He [man] feels afresh in himself for the first time, a product of the process of evolution, perceives that process and reads in it his own making. . . . Lessons from the old sub-human existence enjoined on him showing him at least what to avoid. But ancient trends die hard. . . . He must try to shed from his gene complex some sub-human ingrained elements. . . . There was a time when he nursed the notion that he stood a thing apart, even somewhat after the manner of an Olympian, or of one of the host of heaven. He was wont to think of man and Nature as two contrasted empires."

G. Serban

As a psychiatrist I am interested in any type of research which has potential usefulness for the understanding and treatment of human behavior disorders. I am deeply convinced that animals provide us, if not with all the answers, cer-tainly with some useful models. I am also of the opinion that we are only at the beginning of these studies, and in the future more useful information will come from this field.

The first series of questions I would like to pose is connected with the several levels of discrepancies which exist between animal and human behavior. First, there are differences between the structures of the nervous system in the animals we are studying and the structure of the human nervous system. Whenever pos-sible, one tries to obviate the difficulty by studying monkeys or, if grants permit (and they rarely do in Europe), anthropoid apes. But we do not know if in certain instances some other animals would not be more relevant. My colleagues in the field of nutrition tell me that the pig is the best model in that case. It may be that animals we are not using would provide us with more relevant models in certain cases than the ones we are using for practical reasons.

Next, we must examine whether the structure of the nervous system of the animal is somewhat similar to the structure of the human nervous system (it can happen that the underlying biochemical metabolism is different). This question is of utmost importance, and every psychiatrist with an interest in psycho-pharmacology is acquainted with the uncertainties of the results in man of drugs carefully tested in animals because the metabolism of man is different from the metabolism of the experimental animal. Even if the behavior or the deviation of behavior seems to be identical in an animal and in man, on a higher level we do not know whether it corresponds to the same mechanism. One of the far-reaching discoveries of the ethologist has been that instinctual behavior is started by internal, or external, cues, and until we know the system of cues that belong to a particular species perfectly, we are not entitled to draw any conclusions. This problem has been raised and discussed in the case of animal homosexuality as a model for human homosexual behavior. We do not know whether animal homosexual behavior does in fact have the same mechanism as the human homosexual behavior, because we do not know if it is started by the same system of cues.

Finally, on the highest level, human behavior is strangely determined by psychological superstructure or, if one prefers, in the language of the Pavlovian psychologists, by the existence of a second system of signalization. The difficulty here lies in the fact that we have to search below the observed psychological phenomenon for the underlying elementary behavior common to animal and to man (drive and motivation is an excellent example of that kind of difficulty).

The second problem I would like to pose is connected with the nature of the observed pathological behavior of animals. This pathological behavior, as usually described, has always been produced by manipulating in one way or another the environment of animals by provoking conflicts or social deprivation, modifying the neurological or biochemical mechanism of the brain by inducing a lesion in it, using drugs, and so on. From a human psychiatric point of view, this type of pathological behavioral model corresponds either to organic syndromes or to psychogenic behavioral disorders. As psychiatrists, we are anxious to obtain from the animal psychologists, or more specifically from the ethologists, an animal model representing the endogenous psychosis of man, of both depression and schizophrenia. The nature of the tests used by psychopharmacologists to discover animal antidepressant or antipsychotic action in drugs shows conclusively that we are very far from having animal models of psychotic behavior in man at this time.

A third and last problem is the choice of methods best suited for finding the relevance of animal behavior for psychiatry. Is it better to use psychophysiological methods, experimental psychological methods, or ethological ones?

Some 20 years ago I witnessed very heated discussions between an experimental psychologist and a pathologist. The experimental psychologist was at that time the great man of French psychology, Professor Pierron, telling the dean of ethology in France, Professor Chauvan, that everything in his first paper was absolute nonsense. The only valid study was traditional animal experimental psychology. My impression (as an outsider) is that this type of conflict is not so acute now, since the experimental psychologists recognize the value of ethology and the ethologists are relying more and more on the methods of experimental psychology.

P. Pichot

We are facing at the present time a state characterized by many as a state of crisis—a state of increasing fragmentation of society in all countries, a tendency toward polarization, with various groups striving for their immediate ends without regard for the long-term consequences. I think one of the things that is most impressive during this period, and of particular concern to scientists, is a steady rise of a commitment to irrationality. We see this in many aspects of society today, for example, increased interest in the occult. One only has to go around some afternoon or evening and see the lines waiting to see *The Exorcist* to know how people are attracted to the mystical. Whether one looks upon this notion as either anti-science or anti-intellectual, one of the basic concepts inherent in this development is that man's behavior is not amenable to investigation by rational, systematic, analytic methods, but that man's behavior can only be determined by some arcane metaphysical method of a response to subjective phenomenon, rather than by any objective investigation. As some put it—you can't determine it with the cerebrum, it has to be done with the heart.

This is a trend that can only set us back in our scientific development. There are many reasons for this trend, but time is not available to discuss them all. The contribution of unwise science certainly has contributed to it. In any event, if we are to advance, we must adhere to the scientific or rational model—and not in a simple-minded fashion. The newer school of the history of science—people like F. Kuhn—have emphasized how complex scientific development is. Acceptance of new ideas is not simply a matter of the compelling logic of research. There are social, cultural, and subjective factors. This is true for theories such as those advanced by Copernicus or Galileo. Certainly, for us in psychiatry and the behavioral sciences, one of the major issues is establishing the validity of our field which is now so questioned. The question of the existence of mental illness and whether there are several mental illnesses is a crucial one. As Professor Pichot has already emphasized, this is a time when we have to develop new models. And new models are needed that do not adhere to the very simplistic dichotomy that has characterized previous discussions. On the one hand, we cannot explain behavior, particularly serious aberrations of behavior, simply on an experiential or an environmental model. On the other hand, there are no simple organic or biological explanations that can account for the complexity and diversity of human behavior. What we must develop are new models in which we get a synthesis of both the experiential and the biological. I believe our knowledge is approaching the point where this could be done quite readily. Further, it is incumbent upon us to close the gap between the basic sciences and the applied sciences. We must concern ourselves on the one hand with basic animal models, but on the other hand, strive to close the gap separating the animal and human models. We realize that the medieval notion of separation between the body and the soul has no place now and we must try to synthesize these two aspects—the biological and the psychosocial. We have learned particularly from the animal models that experiences does have its physiological or chemical concomitants, for example, in regard to memory and learning. Certain experiences can have negative consequences. Thus, deprivation may cause deficiency in the development of certain synapses. We have been warned of the danger of a simple vulgar application of animal experimentation to the human. Perhaps we should be proud of it—human behavior is different and more complex than animal behavior. And since human life and human endeavor are so complex, it must be a very complex model.

A. F. Freedman

The subject of this volume is of utmost concern, particularly in these times when the difference between what was accepted as normal and abnormal behavior is minimized by various social groups. Our contemporary world, by shifting its social values and purposes aimlessly and confusing its people, appears contradictory at best.

Advanced technology has helped us to exercise unprecedented control over our environment but not over ourselves. It is demoralizing and frightening to see how successfully mankind is approaching its own self-destruction. If the specter of our population explosion, famine, overcrowding, and pollution is balanced by the terror of nuclear conflict for control of natural resources, then the future of man is indeed bleak.

Basically, this indicates how little we know or understand ourselves. It is of extreme importance to know to what extent our behavior is the result of free will, controlled by our rationality, or, *vice versa,* to what extent our actions are determined by our instinctual needs and justified afterward by logical explanation. Thousands of years of believing our natures to be of divine origin have not helped us in solving the mystery of ourselves. Perhaps, with humility, we should go back to our ancestors, the animals, to find the clues to our puzzling behavior.

S. Kittay

Instinctual and
Environmental Learning

Factors Affecting Responses to Social Separation in Rhesus Monkeys

STEPHEN J. SUOMI

The use of nonhuman primates as subjects for useful models of human psychopathology has attracted increasing interest in recent years. Such use reflects the growing sophistication of primate researchers in identifying parallels between human and nonhuman behavioral patterns. It has been motivated by a desire for information difficult, if not impossible, to obtain easily from human patients.

We have great respect and sympathy for researchers of human psychopathology who are reluctant to accept animal models such as the "reserpinized rat" yet must contend with formidable ethical and methodological problems in their own work. There are some definite advantages for primate research in this area. First, monkeys are far closer anatomically, intellectually, and socially to humans than are rodents, canines, or felines. Second, ethical and methodological problems in monkey research are relatively trivial compared to those encountered in use of human patients as subjects.

Yet studies of psychopathology in monkey subjects have their own limitations. Primate models are useful to human investigators only to the extent that they facilitate identification of etiology, understanding of symptomatology, and development of appropriate therapy for *human*, rather than animal, patients.

STEPHEN J. SUOMI ● University of Wisconsin Primate Laboratory, 22 North Charter Street, Madison, Wisconsin 53706.

Models based upon purported parallels between monkey and man, which in fact do not exist, can only complicate, not clarify, the human picture. Of crucial importance for animal modelers is the need to determine the degree to which cross-species *behavioral* comparisons reflect underlying *mechanisms* which are consistent in similar species.

We nevertheless believe that strong cases can be made for appropriate monkey models of certain basic human psychopathologies. One of these is depression. The data serve as both an effective guide and an effective warning.

Over the past decade multiple research efforts at Wisconsin and at other laboratories have been directed toward study of the consequences of breaking well-established social attachment bonds between or among monkey subjects. A major impetus for such work was the preestablished human studies offered by Spitz and Bowlby and co-workers of human infants and children who had been separated from one or both parents. Many of those children exhibited severe psychopathological symptoms following separation. Spitz labeled the symptoms "anaclitic depression," while Bowlby identified specific components—"protest," "despair," and "detachment"—in his subjects' reactions. Both investigators identified a social etiology, the breaking of a previously established attachment bond, both clearly described the major symptoms, and both provided meaningful data concerning prognosis and therapy.

Harlow and co-workers at Wisconsin, pursuing study of variables affecting mother–infant attachment in rhesus monkey subjects, were impressed by the apparent power of social separation to disrupt human mother–infant interaction and thus attempted to replicate the human findings in their simian subjects. Seay, Hansen, & Harlow (1962) reared infants with their mothers and peers for the first 6 months of life, then physically separated mothers from infants via a Plexiglas barrier for a period of 3 weeks, after which mother and infant were reunited. The results were dramatic. Infants responded immediately to the separation with violent agitation, characterized by vastly increased vocalization, persistent looking at the mother, and hyperactivity. However, these responses were transient, giving way to withdrawal after a few days (Figure 1). Peer play virtually ceased during the separation (Figure 2). Immediately following reunion, mother–infant interactions were above preseparation levels, but these soon declined to baseline. In short, the monkey–infant reaction to separation from the mother was remarkably similar to the human data in terms of etiology, symptomatology, and therapeutic consequences of reunion.

Of course, a single study of mother–infant separation in monkeys can scarcely form the basis for an animal model of human depression. Two problems are immediately apparent. First, not all human mother–infant separations yield a clear-cut anaclitic depression syndrome. For example, Spitz's (1946) classic study reported the above-described reaction in less than $\frac{1}{4}$ of all subjects examined. The common everyday physical separations, which of necessity occur in

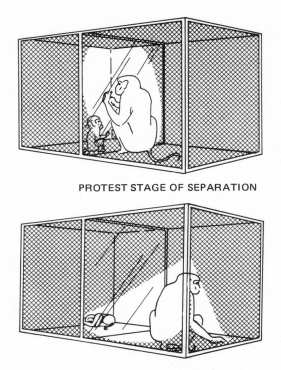

PROTEST STAGE OF SEPARATION

DESPAIR STAGE OF SEPARATION

Figure 1. Stages of infant monkey reaction to separation from mother.

today's living, rarely result in child psychopathology. If they did, we would have an almost universal population of depressed infants. Thus, any useful monkey model of anaclitic depression should contain some reasonable account of specific variables which determine the probability of depressive reaction to separation.

Second, and probably more importantly, a larger question exists as to whether anaclitic depression exhibited by *human* infants is characteristic of the various adult disorders clinically described as depression (see Rie, 1966, for a critique of this assumption). A monkey model must go beyond the mother–infant separation paradigm if it is to be taken seriously as an analog to anything more than human anaclitic depression. The purpose of this report is to present data from recent studies at Wisconsin which address themselves to the above points. Specifically, variables which have been demonstrated to markedly affect the nature of response to social separation by infants will be examined. Next, data regarding depressive responses shown by semiadult rhesus monkeys will be discussed. Finally, some preliminary findings from recent efforts to rehabilitate monkeys following depression-inducing manipulations will be described.

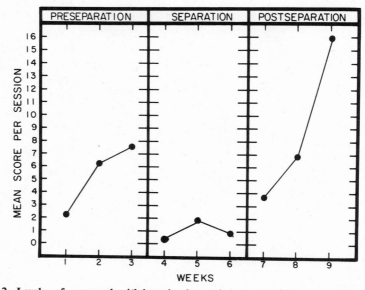

Figure 2. Levels of approach-withdrawal play prior to, during, and after maternal separation.

VARIABLES AFFECTING RESPONSE TO SOCIAL SEPARATION IN MONKEY INFANTS

Although the response to maternal separation described by the original Wisconsin study has been basically replicated by several investigators at different laboratories working with a variety of macaque species (e.g., Seay & Harlow, 1965; Hinde, Spencer-Booth, & Bruce, 1966; Kaufman & Rosenblum, 1967; Schlottmann & Seay, 1972), close examination of the data indicates that the nature and severity of infant response can be affected by several factors. For example, both Hinde and his co-workers (1966, 1970) and Kaufman & Rosenblum (1967, 1969) reported that differences in mother–infant interaction patterns prior to separation could be predictive of differences, either between or within species, in infant responses to separation. Specifically, infants who were most active in maintaining proximity to the mother and who were the recipient of the greatest amount of maternal rejection tended to exhibit the greatest amount of disturbance following separation from mother.

Several subsequent studies performed at Wisconsin also strongly suggested that preseparation variables have considerable influence on response to separation. For example, Suomi, Harlow, and Domek (1970) found that depressive reactions similar to those previously reported to result from mother–infant separation could be reliably induced by repetitive separation of infant monkeys from peers. During each of twenty 4-day peer separations, subjects exhibited biphasic protest–despair reactions, and during reunion periods they showed extraordinary levels of peer-directed activity (Figure 3). A striking cumulative effect of the separations was almost total maturational arrest of behavioral development (Figure 4), despite the fact that the 6-month period during which subjects were repetitively separated is chronologically the period of maximal positive social development in nonseparated subjects similarly reared (Suomi, 1973).

However, additional research (McKinney, Suomi, & Harlow, 1972; Bowden & McKinney, 1972) demonstrated that peer separation per se did not invariably result in depressive psychopathology. Rather, it was found that 2- and 3-year-old feral or mother-reared monkeys living with peers showed only moderate protest when peer-separated and little, if any, evidence of despair. Although simple age differences might have accounted for such differential response to peer separation, preseparation observations suggested that peer relationships in these sub-

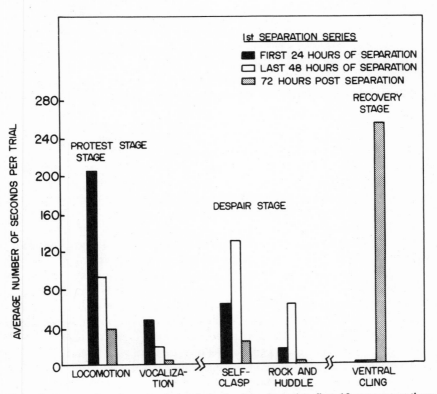

Figure 3. Protest, despair, and reunion behavioral levels during first 12 peer separations.

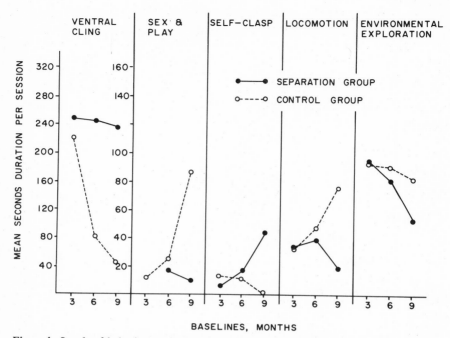

Figure 4. Levels of behaviors at 3, 6, and 9 months among separated and control monkeys.

jects were not nearly as intense as those of the younger peer-reared subjects in the Suomi et al. (1970) study. Several other studies (Suomi, 1971) also revealed a relationship between preseparation behaviors and response to separation. In general, it was found that the nature of a subject's *entire* behavioral repertoire, not simply attachment behaviors, could be predictive of a subject's response to separation from a variety of social environments.

Thus, it seems that, at least for monkeys, social and perhaps nonsocial environmental factors may be correlated with response to subsequent separation, whatever its form. Yet an increasing number of recent studies have reported the importance of the question, who is being separated from whom and how, in determining subject response to separation. For example, in a study involving the previously described procedure of multiple, short-term separations of peers, Suomi (1973) compared the reaction of infants housed in individual cages during their 20 4-day peer separations with the reactions of infants placed in Harlow vertical chambers (illustrated in Figure 5). Both groups showed biphasic responses to all separations and exhibited severe maturational arrest by the the end of the study. However, during reunion periods, chamber-separated subjects showed less peer-directed activity and more depressive self-directed behavior such as self-

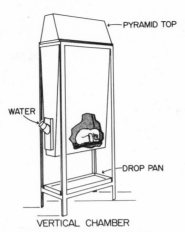

PYRAMID TOP

WATER

DROP PAN

VERTICAL CHAMBER

Figure 5. The Harlow vertical chamber apparatus.

clasping than did their cage-separated counterparts (Figure 6). Suomi pointed out that previous work had demonstrated that chamber confinement per se was associated with subsequent increases in depressive, self-directed behavior, possibly facilitated by the shape of the chamber, and he argued that nature of the physical separation could markedly affect the reactions of similarly reared subjects to separation. Hinde's group has reported a similar finding (Hinde & Davis, 1972). Monkeys separated from not only their mothers but also their familiar social environment exhibited different reactions, both during separation and following reunion, than did monkeys whose mothers were removed from their existing social environment.

Thus, it appears that nature of separation can affect the course of monkey reaction to separation. Other data suggest that not only separation environment but also reunion environment may influence short- and long-term subject behavior. For example, Suomi & Harlow (1975) examined the effects of *sequential* peer separation and/or chambering upon groups of four surrogate-peer-reared monkeys. In two groups, a subject was removed at 4 months of age, placed in a vertical chamber for 28 days, then returned to the group. After 14 days, a second group member was removed, chambered, then returned. This procedure was continued until each group member had been chambered. Members of a third group were sequentially removed as above, but were housed in individual cages rather than vertical chambers during separation. A fourth group remained intact throughout the study.

Most separated subjects exhibited clear-cut protest–despair during the term of removal from their group and, as could be predicted, chambered subjects generally showed more severe reactions than did cage-separated subjects. Of interest,

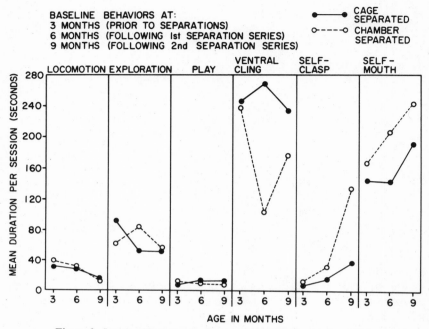

Figure 6. Levels of behavior for cage vs. chamber-separated monkeys.

though, was the gradual and progressive deterioration of group activity as members were systematically removed and returned. Behavior levels did not attain preseparation levels following each reunion, but regressed instead (Figure 7). Predictably, group deterioration was greater among chambered groups, and, interestingly enough, was more pronounced in the chambered group whose most dominant member was removed first.

A more dramatic illustration of the power of postseparation environment upon long-term reaction to separation can be found in Suomi, Collins, & Harlow (1973). Here monkeys were reared with their mothers for 60, 90, or 120 days, then separated permanently. Half the subjects in each age group were then housed individually, while the other half were housed in pairs. Although all subjects reacted to separation with initial protest, only individually housed subjects subsequently exhibited chronic despair-type behavior (Figure 8). With respect to long-term effects, subjects housed in pairs developed a behavior repertoire roughly equivalent to that of well-socialized monkeys of similar age (Figure 9). In contrast, by 6 months of age individually housed subjects behaved like monkeys reared from birth in partial social isolation (Figure 10). Postseparation environment clearly played an enormous role in the psychopathology, or lack of it, shown subsequent to maternal separation in these subjects.

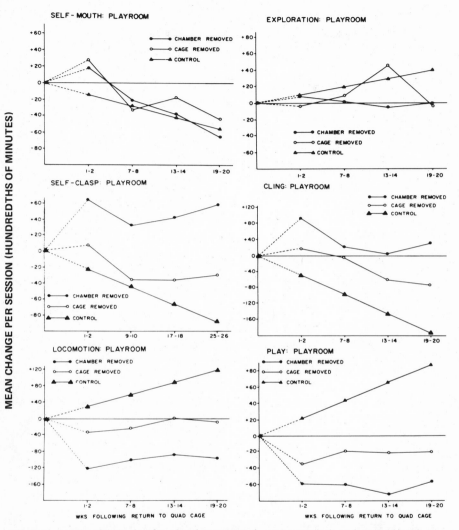

Figure 7. Levels of behavior for chamber-separated, cage-separated, and control subjects.

At this point in the argument for a viable monkey model of depression, certain qualifications seem appropriate. First, the assumption that anaclitic depression is a descriptively useful analog to adult depression is hardly strengthened by the finding that manipulations which consistently produce depression in infants seem to lose their power when juvenile-age subjects are employed. Second, the statement that preseparation variables, nature and severity of separa-

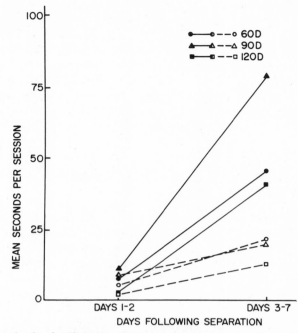

Figure 8. Mean levels of self-clasping following permanent maternal separation. The 60D, 90D, and 120D refers to age in days at time of separation, solid lines are for individually housed subjects while broken lines are for subjects housed in pairs.

tion, and form of postseparation environment are important determinants of reaction to separation may sound impressive, but it is predictively useless unless the variables can be identified and their specific influences described.

One must admit that skepticism is meritorious. Obviously it is inappropriate to proclaim that anaclitic depression is entirely equivalent to all disorders clinically diagnosed as depression. Yet we offer the following data as an argument that perhaps not all analogies ought to be dismissed out of hand. These data were obtained from semiadult monkeys reared from birth in laboratory nuclear family social environments (M. K. Harlow, 1971). Here, as illustrated in Figure 11, monkeys are reared with mothers, fathers, and siblings, and they have access to other similarly structured families.

Ten monkeys, so reared for about 5 years, were removed from their family environments for a period of 120 days. Four were housed with friends, four were housed as pairs with strangers, while the remaining two were housed individually. All subjects showed brief and mild disturbance immediately following separation. The four familiar subjects soon settled back to their customary repertoire of behavior. The two unfamiliar pairs showed totally predictable aggression

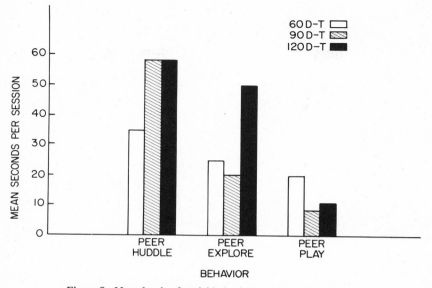

Figure 9. Mean levels of social behaviors, 150–180 days of age.

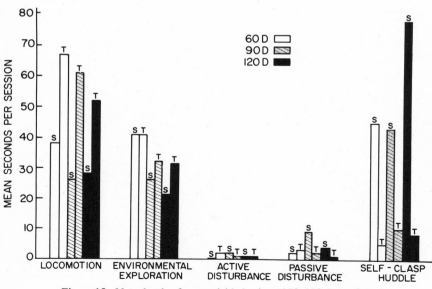

Figure 10. Mean levels of nonsocial behaviors, 150–180 days of age.

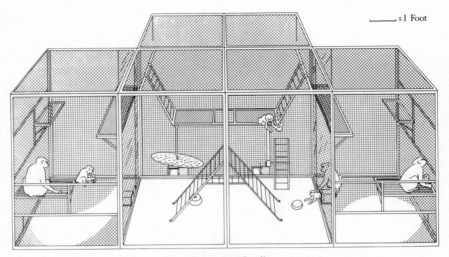

_____ = 1 Foot

Figure 11. The nuclear family apparatus.

upon being introduced to one another, but they, like the familiar group, soon settled down to relatively normal activity. It was not so with the semiadult monkeys housed individually following removal from their families. These animals instead exhibited the classic Bowlby protest–despair reaction to separation from object(s) of attachment. Substantial amounts of self-clasping and huddling behaviors were observed, rather remarkable in light of the subjects' age and rearing history and sharply in contrast to the McKinney findings with preadolescent monkeys. Had these reactions been observed in infant monkeys, they would undoubtedly have been diagnosed as anaclitic depression.

And so, rather than dismiss or blindly accept analogy out of hand, perhaps some sensible conclusions regarding the monkey data can be advanced. First, it seems clear that separation yields responses in monkeys probably *as* variable as the responses it yields in humans. There is no universal reaction to social separation in monkeys. Second, some separations in monkeys yield responses tantalizingly similar to those of humans diagnosed as clinically depressed. But probably more importantly, knowledge of certain variables can facilitate prediction of monkey response to any given form of separation. One can talk of preseparation variables and postseparation variables, but perhaps a more meaningful conception can be found in answer to the following questions: What is lost by each subject? To what degree are the losses replaced in the postseparation environment? And, what are the limitations of the postseparation environment on the subjects' subsequent social activity?

If one examines the data from all of the above-mentioned monkey studies in search of answers to these questions, certain consistencies with respect to monkey reaction to separation emerge from study to study. For example, in the above data it is clear that if separation did not invoke loss of the opportunity to interact physically with those who previously played a major role in one's social activity, reaction was mild. Kaufman and Rosenblum's (1969) bonnet infants lost passive part-time mothers; the McKinney adolescents lost age-mates to whom, on the basis of preseparation behaviors, they could scarcely have been described as strongly attached; and in these cases, depression did not emerge following separation. On the other hand, most subjects who lost what, by preseparation behaviors could be described as strong objects of attachment, exhibited severe, depressive-like reactions to separation. Thus, Suomi's infants who seemed as strongly attached to each other, on the basis of preseparation behavioral levels, as virtually any mother–infant pair, exhibited classic anaclitic depression after *each* of their 20 peer separations. Hinde's data are equally interesting: Infants who were most responsible for maintaining mother–infant ties were those who showed the most severe and persistent reaction to maternal separation.

In cases where the "losses" were at least partially replaced in the postseparation environment, severe depression was a rare outcome. Kaufman and Rosenblum's bonnet infants found mothering "aunts" in their postseparation environment. There was no depression. In almost all studies where reunion took place in the same social environment as separation, overtly depressive behavior, e.g., self-clasping, disappeared rapidly. But where the postseparation environment provided little in the way of substitution of loss produced by separation, depression was a common result. For example, infants housed in pairs following maternal separation (Suomi, Collins, & Harlow, 1973) did not exhibit despair behaviors, while infants housed individually developed deep depressions. Subjects placed in vertical chambers during repetitive peer separations exhibited higher levels of withdrawal behavior during reunion than did subjects housed in cages during comparable separations and reunions (Suomi, 1973). The data are consistent: The less the postseparation environment provided in the way of substitution for the lost attachment bond, the more severe was the response to separation.

Consideration of the limitations of the eventual postseparation environment entails examination of long-term data. For example, Hinde and Davies (1972) found long-term effects of maternal separation to be more severe among infants whose mothers were most adversely affected by the separation experience. More strikingly, Suomi and co-workers (1970, 1973) returned infants to an environment containing infants affected as adversely as themselves, and the result was maturational arrest. A social environment of progressively deteriorating group activity (Suomi & Harlow, 1975) was obviously not conducive to subject development of normal social capability. An even poorer social environment was provided the individually housed subjects of Suomi, Collins, & Harlow (1973), who by 6 months of age exhibited behavioral repertoires similar to those of equal-aged partial isolate-reared subjects (Suomi, Harlow, & Kimball, 1971), despite their 60, 90, or 120 days of mother-peer rearing.

The potential utility of the above approach to interpretation of the monkey data can perhaps be illustrated by reexamination of the nuclear family separation data. All 10 subjects were physically removed from the environment—mothers, fathers, siblings, peers, and other adults—in which they had spent almost 5 years of their lives. Four subjects were housed with life-long friends from their previous environment, and they displayed a rapid return to relatively normal behavior. Four subjects were housed with a friend and two strangers. These subjects spent a predictable amount of time adjusting to their new, different, but probably challenging social environment. But two of the subjects were housed individually. Their environment contained no family members, no friends with whom bonds had been built up over almost 5 years, not even any strangers. In short, their postseparation environment provided absolutely no substitutes for the loss occasioned by separation. Further, it offered the subjects very little opportunity, social or nonsocial, for constructive activity. These subjects responded to separation with depression. Given appropriate social conditions, then, even semiadult socially competent and sophisticated monkeys can be made to exhibit psychopathology.

THERAPY

What is the nature of the preseparation environment? What is lost by separation? And, what are the limitations of the environment subsequent to separation and/or reunion? Consideration of these questions appears to help account for the various reactions to separation described in the literature. Consideration of answers to these questions could also conceivably influence determination of procedures aimed at rehabilitation of depression.

We at Wisconsin have been engaged for almost 5 years in research involving rehabilitation of experimentally induced psychopathology in monkeys. Until now we have focused upon rehabilitation of monkeys reared in total social isolation. Total social isolation involves raising animals in chambers and without any social contact. The chambers are diffusely illuminated and not sound shielded so that the confounding of social deprivation with sensory deprivation is minimized (see Harlow & Harlow, 1962). Such isolation produces severe psychopathology, although most likely a direct human analog does not exist, as fortunately, total social isolation-rearing of human infants is hardly a widespread practice. Nevertheless, our most successful strategies for rehabilitation have entailed considerations not unlike the questions posed above. What is lost by social isolation from birth? One must consider that monkeys reared in isolation have no chance to develop contact–comfort attachment to a mother-figure, and possibly more importantly, have no chance to develop adult social capabilities through play interactions with peers. Instead, they develop stereotypic and self-directed behavior nonadaptive in social situations. What should an adequate postisolation environment provide? We believed that it should initially provide opportunity to develop contact security while simultaneously rendering self-directed activity incompatible with social contact-oriented activity. Subsequently it should provide the opportunity for gradual development of social activities such as play. On this basis, we chose to introduce our isolates to socially normal "therapists" who were younger chronologically. Being younger, the therapists were not social threats. Furthermore they were at an age where play is just starting and, step by step they could lead the isolates up to the full vigor and vivacity of social play. The results were impressive: 6- and 12-month social isolates achieved age-normative levels of social and nonsocial activity characteristic of well-socialized monkeys (Suomi & Harlow, 1972; Harlow & Novak, 1973; Cummins, 1973).

Similarly, we have taken into account our questions regarding reaction to separation when designing studies for the purpose of rehabilitating monkey depression. For example, we are presently providing monkey therapy to infants subjected to repetitive peer separations. As described above (Suomi, Harlow, & Domek, 1970), monkeys reared together, without mothers, from birth, then separated repeatedly from each other show protest and despair with each separation, excessive clinging and little play during reunions, and pronounced maturational arrest. What is lost during separation? Subjects are denied physical access to the only social objects with whom they have ever had an opportunity to interact. What does the separation environment provide? No opportunity to interact physically with any social agents is permitted—depression results. What are the limitations of the reunion environment? Reunited subjects interact only with clinging, nonplaying peers—maturational arrest results.

Given these considerations, it was clear to us that the clinging, socially unsophisticated type of monkey therapist used successfully for rehabilitating social isolates would be inappropriate for repetitively separated subjects. Such therapists might alleviate despair during separations but they would do little to prevent maturational arrest. Instead, we chose to employ socially active and playful age-mates as therapists, monkeys who could be expected to initiate play invitations consistently when introduced to any peer.

In a study recently completed, a group of four monkeys were reared together without mothers from birth, for the first 3 months of life. They were then subjected to twenty 4-day peer separations over the next 6 months. For the first 12 separations they received no therapy, and the results were predictable—protest and despair during separation, intense peer-clinging during reunion, and maturational arrest. At this point the therapists were introduced—surrogate-peer-reared age-mates who had been given extensive opportunities to interact with other monkeys in a variety of social environments. For the final eight separations, subjects were allowed to interact with therapists for 2 hours daily during both periods of separation and periods of reunion.

Preliminary results have been most encouraging. The data examined to date suggest that the severity of protest and despair was much reduced with increased exposure to the therapists. At present the subjects show high levels of play and low levels of clinging and withdrawal during therapy sessions. We will have to wait until the completion of the study to make a complete evaluation of the degree of maturational arrest, but the results as of now are promising.

PERSPECTIVES

At present, then, we have data relevant to two forms of psychopathology in monkeys—that which results from social isolation from birth and that which results from the breaking of attachment bonds under certain conditions. They differ significantly from each other in origin, in distinctive behavioral concomitants, and in procedures which successfully reverse the psychopathologies. We believe that one probably does not have a common human analog, while the other most likely does, particularly since it was initially based almost exclusively on human data. These findings convince us that monkeys are definitely capable of developing syndromes of psychopathology and, given the proper precipitating conditions, they probably are capable of developing syndromes analogous to many of those commonly diagnosed in man.

The degree to which monkey syndromes mirror human syndromes is without question dependent on the degree to which the obvious disparity between the behavioral and cognitive capabilities of monkey and man is important for any specific syndrome. For example, if a particular human syndrome can be "explained" exclusively in terms of form of social attachment bonds and those forms exist in monkeys, then it is probable that monkeys can exhibit a comparable syndrome, given their behavioral and cognitive limitations. If a particular human syndrome can be "explained" exclusively in terms of a genetic defect in catecholamine synthesis and a similar deficit is present in some monkeys, then it is probably that a monkey model will be viable. If, on the other hand, a particular human syndrome can be "explained" exclusively in terms of processing of verbal information, then it is unlikely that a monkey model exists, for monkeys are behaviorally and probably cognitively incapable of verbal fluency and understanding.

Much of our primate research at Wisconsin is directed toward assessment of the physiological, social and cognitive capability of rhesus monkeys. We are not unaware of human data—quite frankly, they tend to bias our research efforts (Harlow, Gluck, & Suomi, 1972). For example, we are presently investigating the possibility that there may exist monkey analogs to sociopathology and to schizophrenia. The analogs may exist and then again they may not—the data-to-date are hardly definitive. If, in fact, they become definitive, we can only hope that they will be viewed with the considerations that presently are afforded definitive human data.

ACKNOWLEDGMENTS

This research was supported by National Institute of Mental Health grants MH-11894, MH-18070, MH-47025, and by the Grant Foundation.

REFERENCES

Bowden, D. M., & McKinney, W. T. Behavioral effects of peer separation, isolation, and reunion on adolescent male rhesus monkeys. *Developmental Psychobiology*, 1972, *5*, 353–362.

Cummins, M. S. Behavioral stability of rhesus monkeys following differential rearing. Unpublished M. A. thesis, University of Wisconsin, 1973.

Harlow, H. F., Gluck, J. P., & Suomi, S. J. Generalization of behavioral data between nonhuman and human animals. *American Psychologist*, 1972, *27*, 709–716.

Harlow, H. F., & Harlow, M. K. The effect of rearing conditions on behavior. *Bulletin of the Menninger Clinic*, 1962, *26*, 213–224.

Harlow, H. F., & Novak, M. A. Psychopathological perspectives. *Perspectives in Biology and Medicine*, 1973, *16*, 461–478.

Harlow, M. K. Nuclear family apparatus. *Behavior Research Methods and Instrumentation*, 1971, *3*, 301–304.

Hinde, R. A., & Davies, L. Removing infant rhesus from mother for 13 days compared with removing mother from infant. *Journal of Child Psychology and Psychiatry*, 1972, *13*, 227–237.

Hinde, R. A., & Spencer-Booth, Y. Individual differences in the responses of rhesus monkeys to a period of separation from their mothers. *Journal of Child Psychology and Psychiatry*, 1970, *11*, 159–176.

Hinde, R. A., Spencer-Booth, Y., & Bruce, M. Effects of 6-day maternal deprivation on rhesus monkey infants. *Nature*, 1966, *210*, 1021–1033.

Kaufman, I. C., & Rosenblum, L. A. The reaction to separation in infant monkeys: Anaclitic depression and conservation-withdrawal. *Psychosomatic Medicine*, 1967, *29*, 648–675.

Kaufman, I. C., & Rosenblum, L. A. The waning of the mother–infant bond in two species of macaque. In B. M. Foss (Ed.), *Determinants of infant behavior*, Vol. 4. London: Methuen, 1969.

McKinney, W. T., Suomi, S. J., & Harlow, H. F. Repetitive peer separations of juvenile-age rhesus monkeys. *Archives of General Psychiatry*, 1972, *27*, 200–204.

Rie, H. E. Depression in childhood. *Journal of the American Academy of Child Psychiatry*, 1966, *5*, 653–685.

Schlottmann, R. S., & Seay, B. Mother–infant separation in the Java monkey (*Macaca irus*). *Journal of Comparative and Physiological Psychology*, 1972, *79*, 334–340.

Seay, B., Hansen, E. W., & Harlow, H. F. Mother–infant separation in monkeys. *Journal of Child Psychology and Psychiatry*, 1962, *3*, 123–132.

Seay, B., & Harlow, H. F. Maternal separation in the rhesus monkey. *Journal of Nervous and Mental Disease*, 1965, *140*, 434–441.

Spitz, R. A. Anaclitic depression. *Psychoanalytic Study of the Child*, 1946, *2*, 313–342.

Suomi, S. J. Experimental production of depressive behavior in young monkeys. Unpublished doctoral dissertation, University of Wisconsin, 1971.

Suomi, S. J. Repetitive peer separation of young monkeys: Effects of vertical chamber confinement during separation. *Journal of Abnormal Psychology*, 1973, *83*, 1–10.

Suomi, S. J., Collins, M. L., & Harlow, H. F. Effects of permanent separation from mother on infant monkeys. *Developmental Psychology*, 1973, *9*, 376–384.

Suomi, S. J., & Harlow, H. F. Social rehabilitation of isolate-reared monkeys. *Developmental Psychology*, 1972, *6*, 487–496.

Suomi, S. J., & Harlow, H. F. Effects of differential removal from group on social development of rhesus monkeys. *Journal of Child Psychology and Psychiatry*, 1975, *16*, 149–164.

Suomi, S. J., Harlow, H. F., & Domek, C. J. Effect of repetitive infant–infant separation of young monkeys. *Journal of Abnormal Psychology*, 1970, *76*, 161–172.

Suomi, S. J., Harlow, H. F., & Kimball, S. D. Behavioral effects of prolonged partial social isolation in the rhesus monkey. *Psychological Reports*, 1971, *29*, 1171–1177.

Human Personality Development in an Ethological Light

JOHN BOWLBY

In this chapter I describe a problem in human psychopathology and some of the many ways in which studies of animals are contributing to its solution.

By the early 1950's there was considerable evidence that lack of continuous loving care from a mother-figure could result in a child growing up with serious personality defects, e.g., sociopathy (Goldfarb, 1943, Bowlby, 1944); that prolonged stay in an institution could result in retardation (Skeels & Dye, 1938-39, Spitz, 1945); and that between the ages of 12 months and four years even a week or two of being cared for in a strange place by a succession of strange people could result in a child showing intense emotional responses of protest and despair and, on reunion with his mother, a curious emotional detachment from her, followed usually by intense clinging and fear of further separation (Robertson & Bowlby, 1952). At that time the evidence was fragmentary and the role of different variables insufficiently understood. Nevertheless, in view of its relevance to psychiatric etiology and prevention, the field appeared a promising one for systematic research. (For reviews of evidence, see Bowlby, 1951, Ainsworth, 1962, Rutter, 1972).

Although as a psychiatrist my concern has been to understand the long-term effects on personality development of disrupted mother–child relations occurring during the early years of life, and the variables that determine favorable or

JOHN BOWLBY • Tavistock Institute of Human Relations, School of Family Psychiatry and Community Mental Health, Tavistock Centre, London, England.

unfavorable outcome, as a research strategy my group decided in 1952 to concentrate initially on the short-term responses. These we came increasingly to regard as the prototypes of responses that, when seen in acute and chronic form in older individuals and out of family context, are habitually labeled as psychiatric symptoms. Thus, for example, when the family situation is ignored, it becomes easy to label despair as depression, detachment as psychopathic lack of affect, protest as hysterical, clinging and fear of separation as overdependency. While further empirical studies of the effects on young children of short separations in different settings were undertaken by my colleagues, notably James Robertson (Robertson & Robertson, 1971) and Christoph Heinicke (Heinicke & Westheimer, 1966), I myself began to consider the theoretical framework within which the responses could best be conceptualized. A particular problem was to understand the nature of the bond between child and mother, disruption of which commonly causes such great distress.

Up to that time my thinking, like that of most others working in the field, had been cast mainly in a psychoanalytic mold because, despite evident deficiencies, that was the only tradition in psychiatry or psychology that had attempted to deal with the problems concerned. In that tradition, as in learning theory, it was then confidently assumed that the child's tie to his mother was the expression of a secondary drive learned instrumentally through the mother satisfying his primary drive of hunger. That assumption seemed to me implausible because it failed to account for the specificity and persistence of a child's tie to the particular woman who mothered him; and it failed also to do justice to the complexity of mother–child interaction, the importance of which for understanding personality development clinical experience in family psychiatry was daily impressing upon me. But if the bond was not a derivative of feeding, how else could it be understood?

It happens that the early 1950's saw the first publications in English of major works by Lorenz (1950, 1952) and Tinbergen (1951). In addition to finding ethological writings of great intrinsic interest, I was struck by the overlap of subject matter between ethology and psychoanalysis. Both were interested in species-specific behavior, especially that mediating affectional bonds between parents and young and between mates; both were interested in the effect of early experience on later development, in conflict arising in social situations, in redirection of behavior, and in displacement activities. This overlap, combined with the sophistication of the ethologists' scientific approach, seemed to promise great things. Thereafter I committed myself to exploring how far ethological principles could help an understanding of the clinical problems with which I was concerned.

In the mid 1950's Harlow, inspired by Spitz's studies of human infants (Spitz, 1945), but himself working in a quite different tradition, began his experimental studies of rearing rhesus monkeys in a severely impoverished environment. He was soon followed by Hinde who, using the same species but an entirely different research strategy (Hinde & Spencer-Booth, 1968), attempted to replicate the situation of a human child left at home for a week or two while his mother was away. Meanwhile, Mary Ainsworth, a clinical psychologist fast acquiring an ethological perspective, was pioneering naturalistic and, later, experimental studies of human infants and their mothers. As a result of these and related research programs, the past 20 years have seen the development of a combined research operation by behavioral scientists from a number of distinct traditions, directed at understanding the place in human personality development of a child's relation to his mother-figure and the effects on that relation of the many variables that influence it, especially periods of separation from the mother and deprivation of maternal care.[1] In this operation there has been a most rewarding two-way process of mutual influence in which those of us working with humans have applied concepts and methods derived from colleagues working with animals, while they in turn have focused experimental attention on problems and variables identified by clinicians.

Reflecting on the possible ways that animal studies can contribute to an understanding of human behavior Hinde (1972) has listed four ways in which animal studies can be of value:

(a) By developing methods of observation, recording and analysis that may prove applicable to research on humans.

(b) By the experimental study of analogous problems in animal species, often using methods that would be ethically impermissible in the case of man. Although the results of such experiments cannot be extrapolated, they call attention to processes that might occur in man, and to variables that might be relevant, and that can then be looked for in humans.

(c) By developing theoretical concepts and perspectives that may prove applicable.

(d) By calling into question explanations of human behavior that draw on concepts applicable only to humans whenever the responses seen in a nonhuman species are found to resemble closely those seen in humans.

Research on human problems arising from separation and deprivation has, I believe, benefited in all these four ways. In addition, I would add the following two, both of which are amplifications of (b) above:

(e) By focusing attention on certain typical responses in nonhuman species, attention may be called to their being typical also of humans.

---

[1] It is recognized that, in the case of humans, separation from the father and deprivation of paternal care can also be of great importance. Since the role of the father in the development of a nonhuman primate is very different from what it is in the case of the human child, discussion of this variable is omitted from this paper.

(f) By demonstrating experimentally that certain sequences of development undoubtedly occur in other species, hypotheses already advanced that they occur in humans, which may hitherto have been met with incredulity and hostility, become more acceptable; thereafter, much more intelligent debate becomes possible.

Since my own work has consisted mainly of utilizing ethological principles to reformulate clinical theory, it is to that aspect of the enterprise that the remainder of this paper is directed.

In the development of ethology—and also in my own thinking—the concept of imprinting has been seminal, even though Lorenz's original views have had to be substantially modified. Whether a version of the processes now believed responsible for imprinting occurs in man is unknown because the necessary experiments are ethically impermissible (although such data as we do have are consistent with that idea). My reason for referring to imprinting is a different one. The very fact that such processes occur in many vertebrate species serves to draw attention to a number of principles governing the development and organization of behavior which were omitted in the older traditions of theorizing but which appear to be of the greatest relevance for understanding clinical problems. From among these many principles I concentrate on two, both of which have far-reaching implications.

First, the phenomenon of imprinting demonstrates that, at least in some species, persisting bonds between individuals can develop without any of the traditionally considered physiological drives having to be satisfied, and even in spite of punishment. This frees us from having to make the assumption that a human infant's bond to his mother can be conceived as developing only as a result of his hunger being satisfied (Bowlby, 1958). The probable relevance of that conclusion to the human case was greatly increased when Harlow demonstrated the role of contact–comfort in the development of a rhesus monkey's attachment to a dummy mother (Harlow, 1958).

Secondly, the phenomenon of imprinting draws attention to the termination of behavior and not merely to its activation. When the preferred object is absent, the young animal searches for it and gives distress calls. When visual or auditory contact is regained, the young animal *ceases* to search and *ceases* to give distress calls. Such a sequence is not easily dealt with by an energy theory of motivation. What is the energy that is seeking discharge in searching? How are we to conceive its having been discharged when the preferred object is no more than seen or heard? Why should mere proximity or contact give all the appearances of bringing deep satisfaction?

Reflection on these behavioral sequences, which are just as evident in human children after the age of six months as they are in the young of other species, brings the realization that the sequences are much more readily understood in terms of control theory than of energy theory.

Among the features of control theory especially valuable to an understanding of animal and human behavior are the sharp distinctions it enables us to draw between causation, set-goal, and function, matters about which there have been and still are immense confusions. These distinctions can be illustrated by reference to the control of body temperature. In that case the set-goal of the system is the stable temperature it is geared to maintain. Its biological function is the contribution to a species survival that maintenance of such body temperature makes, namely efficient metabolism despite wide variation of ambient temperature. Causal factors in the system include all those signals that either activate or inactivate the physiological and behavioral processes concerned. Insofar as the system is constructed so that a set-goal is achieved, it is purposive. Insofar as the causal factors are related to the system's purpose only indirectly, through the system's structure, their role can be studied scientifically and without invoking teleological ghosts. The presence of the control system as part of the make-up of an organism is to be explained in terms of evolution theory: organisms so endowed survive better and leave more progeny than do organisms not so endowed (or less well endowed). Thus all the behavior that, in the past, biologists and psychoanalysts have attributed to instinct can in principle be conceived as resulting from the activation and termination of behavioral systems that usually, but not always, develop along lines characteristic of the species.

Another feature of control systems that it is important to note is that they can operate effectively only within certain environmental limits. For example, the system responsible for maintaining body temperature is effective only provided the ambient temperature lies within certain limits. Outside those limits it fails. Thus, throughout the living world, organism and environment cannot properly be thought of as separate things. Organism is only intelligible in terms of the environment to which, through the processes of evolution, it is adapted and cannot usefully be studied except in relation to that environment.

Once we come to consider whether and in what ways ethological principles and control theory can cast light on our particular problem, observations fall readily into place. The evident tendency for a human child or an immature animal to maintain proximity to his mother-figure, and subsequently to other familiar adults, can be regarded as a form of homeostasis in which distance from and/or accessibility to the mother-figure are the parameters that must be kept within certain limits. Effector processes can be considered as all those forms of behavior that increase or maintain proximity, for example, locomotion toward the mother and clinging to her, and also those that encourage the mother to maintain proximity and provide care, such as crying, smiling, and other forms of greeting. Causal factors can be considered as, on the one hand, all those that activate the different forms of attachment behavior and, on the other, all those that terminate them. In the former category are, among many others, pain, fatigue, and anything which alarms a child; in the latter, visual or auditory presence of the mother and especially touching and being held by her. Clearly, to account for all that we know about attachment behavior in its minute-to-minute changes, in its ontogeny, and in the variations it shows from individual to individual requires us to postulate a fairly complex type of control system, the details of which have to be specified. Nevertheless, it is already clear that control theory is, in principle, well suited to explain the data and is far more promising for the purpose than is an energy-type theory (Bowlby, 1969).[2]

It has long been evident that the forms taken by a child's tie to his mother undergo change as he grows up, but that certain of its features remain unchanged. For example, although a child of three years is much more able than a two-year-old to explore away from his mother-figure for a few hours, confident that she will return, he is likely to be very upset should she not do so on time. The same is true, moreover, of a school child and of an adolescent, though as a child grows older the periods he can confidently bridge become longer and longer, and, in addition, he becomes more able to disguise both from himself and from others when he is upset because support is not forthcoming. Similar behavior and feeling also occur in adults, a fact that theory always tends to ignore. Thus there can be no doubt that proximity keeping to certain discriminated individuals is a basic type of human behavior which has its own natural history and must be regarded as a healthy and important part of human nature. It is also clear that it is a type of behavior that in atypical environments, of which an impersonal institution is an extreme case, can develop in one of a number of aberrant ways that are important clinically. This conclusion has been made far more acceptable to psychiatrists and psychologists since Harlow's demonstration that analogous forms of aberrant development also occur in rhesus monkeys brought up in a socially impoverished environment (Harlow & Zimmermann, 1959).

Once attachment behavior is conceived of as an integral part of human nature, the question arises what its biological function may be. The answer arrived at, after scrutinizing evidence from studies of hunting-and-gathering tribes and from studies of other ground living primates in the wild, is protection. By keeping in proximity to others, especially those who are stronger and more experienced, an animal is less likely to come to harm, especially harm from predators, than if it is alone. Even in present-day Western societies, in many of which predators have ceased to be a threat, keeping proximity to an adult promotes safety, as an examination of the incidence of traffic accidents to children who are accompanied or unaccompanied by an adult suggests. Whereas to postulate protection from predators as the function of attachment behavior is treated by

---

[2] A conclusion similar to and consistent with mine has been reached independently by Peter-freund (1971). He presents a cogent criticism of traditional psychoanalytic metapsychology and proposes that the clinical phenomena to which Freud drew attention be reformulated in terms of theory derived from recent studies of information processing and control systems.

ethologists as an almost painfully obvious probability, to clinicians and psychologists it comes as a surprise and is often greeted with incredulity.

Thus all the forms of behavior that mediate a child's tie to his mother, and subsequently to his father and other familiar adults, and to which the inclusive term attachment behavior can be applied, are postulated to have a biological status just as fundamental as eating behavior, with its function of nutrition, and as sexual behavior, with its function of reproduction. Since the function of each of these classes of behavior is distinct and of equal importance for the survival of a species, it is necessary to study the development of each in its own right, including the conditions that make for a healthy or pathological outcome. This provides a new approach to a problem with which psychoanalysts have long been wrestling, namely the frequency with which faulty development in one of these classes of behavior is associated with faulty development in one of or both the other two. Whereas in the past it has usually been assumed that the initial fault occurs in either the feeding or the sexual fields, the new approach frees us to consider whether, perhaps, many such problems originate in the field of attachment behavior. Much evidence, indeed, suggests that they do.

Once the protective function of attachment behavior is agreed upon as likely, a clinician's whole perspective changes. For example, it becomes not too difficult to see how a human's fear of separation and loss is related to his tendency to fear a number of other situations such as strange people and places, darkness, animals, and sudden movement. While Freud (1926), searching for the origins of pathological anxiety, was keenly aware that a child is afraid of separation from his mother-figure and also of all these other situations, he found himself extremely puzzled why that should be so, because none of these situations is intrinsically dangerous. From Freud onward, indeed, psychoanalysts have wrestled with this problem of "unrealistic" anxiety and have advanced a host of different theories, some extremely recondite, to account for it. Psychiatrists working in other traditions have been equally perplexed (Lewis, 1967). When looked at in an ethological perspective, however, a simple solution lies at hand. First, it is recognized that man is not the only species that is apt to show fear in situations that do not carry a high risk of danger. Second, it becomes evident that each of these naturally fear-arousing situations, including separation, have in common that they are associated with an increased risk of danger. Although in each of such situations the *absolute* risk may be quite low, because we have only one life there are great advantages for an animal to be programmed, almost from birth, to respond to each of these situations of *increased* risk with fear. Empirically we find that a specially intense fear is aroused when two or more of these situations are present together, a finding consistent with current theories of information processing and decision making (Broadbent, 1973).

Seen in this light, a human's fear of being separated from a familiar companion can be regarded as a normal and healthy reaction which we share with other mammals and which calls for no specifically human explanation (Bowlby, 1973).

The clinical problem then becomes one of trying to understand why neurotic patients should so often be prone to fear any or all of these situations (and also many others) more readily and more intensely than do other individuals of the same age and sex. Here again, animal studies have given valuable clues, mainly by drawing attention to the relevance of the theory to certain human responses which, although obvious, have been overlooked by clinical theorists.

First came Harlow's studies that showed how a rhesus infant uses his mother-figure (even a dummy) as a secure base from which to explore and to which to retreat when afraid (Harlow & Zimmermann, 1959). Next came Hinde's demonstration that when a monkey is isolated from its companions it shows much more intense fear of a situation than it does when its companions are present (Rowell & Hinde, 1963).

The concept of the secure base has become central both to Ainsworth's empirical studies of children and mothers (Ainsworth, 1967, Ainsworth, Bell & Stayton, 1974) and also to the theories of personality development that she and I are advancing. In particular, it enables us to interpret the results of a large array of studies published in recent years which show a consistent correlation between certain patterns of personality development and certain patterns of family experience during infancy, childhood, and adolescence. In brief, evidence shows that individuals who grow up most stably self-reliant and capable of cooperation with others are those who throughout these years have had parent figures who gave them affection and encouragement and have provided unfailing support in times of difficulty or stress. Conversely, those who show a lack of self-reliance or a tense self-reliance that breaks under stress, and difficulty in cooperating with others, are those who have not had such support: such individuals are likely to develop neurotic symptoms, especially anxiety and depression and/or personality disorders.

Three main classes of experience are found in the histories of those who grow up deficient in stable self-reliance and capacity to co-operate:

(i) Prolonged absence of a mother-figure, changes of mother-figure, or repeated periods of separation from mother-figure.

(ii) Frequent rejection by mother-figure.

(iii) Threats of abandonment by a mother-figure, including threats to commit suicide.

Experiences of deprivation, separation, and rejection can all occur in monkeys; and their adverse effects are now amply demonstrated. Experiences classified under (iii), however, are confined to humans because they entail a conditional clause: "if you do such and such, or don't do such and such, then I shall abandon you." I believe that a principal reason why the crucial part played by such threats in the genesis of pathological anxiety has been neglected by theorists is that the effects of threats are intelligible only when there is a clear recognition that intense anxiety and distress are the natural responses to being abandoned.

To the psychotherapist, what is particularly striking is the way that whatever responses from parent figures an individual has learned to expect during his infancy, childhood, and adolescence he often continues to expect from other human beings during later life, even when his subsequent experiences are entirely different to his earlier ones. Frequently he is partly or even totally unaware that his perceptions, feeling, and behavior are governed by such obsolete expectations; or, if aware of the expectations, he may be unaware how they originated. To understand why expectations that develop during early years so frequently persist virtually unchanged throughout later life, despite much falsifying experience, requires us to invoke principles of cognitive psychology, at least some of which are likely to be specific to humans.

From what I have said there is no need to emphasize how immensely valuable animal studies have been in helping to elucidate the effects on human beings of disruptions of affectional bonds. In the first place, because animal studies have demonstrated that in many respects infants of a nonhuman species respond to the rupture of an affectional bond in ways remarkably similar to those of human infants, we are now able to see certain emotionally laden responses of humans—ones, moreover, that appear to be of the greatest significance to psychiatry—in a broad and comparative light. In the second place, because animal studies have demonstrated by means of experiment the effects on these responses of a number of variables that operate before, during, or after a separation, our confidence that these variables are indeed of the importance we had suspected in the human case is much strengthened. Above all, animal studies have enabled us to develop reliable methods for the study of nonverbal components in the relationship between one human being and another, and to develop, also, ways of conceptualizing the findings that not only do justice to the humanity of man but can be used to generate hypotheses that can then be tested in the usual way against further empirical data. If these claims prove sound, scientifically based progress is assured.

REFERENCES

Ainsworth, M. D. S. The effects of maternal deprivation: a review of findings and controversy in the context of research strategy. In *Deprivation of maternal care: a reassessment of its effects*. Public Health Papers No. 14. Geneva: WHO, 1962.

Ainsworth, M. D. S. *Infancy in Uganda: infant care and the growth of attachment*. Baltimore, Md: The Johns Hopkins Press, 1967.

Ainsworth, M. D. S., Bell., S. M., & Stayton, D. J. Infant-mother attachment and social development: socialization as a product of reciprocal responsiveness to signals. In M. Richards (Ed.), *The integration of a child into a social world*, Cambridge: Cambridge University Press, 1974.

Bowlby, J. Forty four juvenile thieves: their characters and home life *International Journal of Psychoanalysis*, 1944, *25*, 19-53, 107-128.

Bowlby, J. *Maternal care and mental health*. Geneva: WHO; London: HMSO; New York: Columbia Univ. Press; 1951. Reprinted, New York: Schocken Books, 1966.

Bowlby, J. The nature of the child's tie to his mother, *International Journal of Psychoanalysis*, 1958, *39*, 350-373.

Bowlby, J. *Attachment and loss* (Vol. 1). London: Hogarth Press; New York: Basic Books, 1969.

Bowlby, J. *Attachment and loss* (Vol. 2). London: Hogarth Press; New York: Basic Books, 1973.

Broadbent, D. E. *In defence of empirical psychology*. London: Methuen 1973.

Freud, S. Inhibitions, Symptoms and Anxiety. Standard Edition Vol. 20. London: Hogarth Press, 1926.

Goldfarb, W. The effects of early institutional care on adolescent personality. *Journal of Experimental Education*, 1943, *12*, 106.

Harlow, H. F. The nature of love. *American Psychologist*, 1958, *13*. 673-685.

Harlow, H. F., & Zimmermann, R. R. Affectional responses in the infant monkey. *Science*, 1959, *130*: 421.

Heinicke C., & Westheimer, I. J. *Brief separations*, New York: International Universities Press; London: Longmans, 1966.

Hinde, R. A. (1972) *Social behaviour and its development in sub-human primates*, Eugene, Oregon: Oregon State System of Higher Education.

Hinde, R. A., & Spencer-Booth, Y. The study of mother–infant interaction in captive group-living rhesus monkeys. *Proceedings of the Royal Society, Series B*, 1968, *169*, 177-201.

Lewis, A. Problems presented by the ambiguous word "Anxiety" as used in psychopathology. *Israel Annals of Psychiatry and Related Disciplines*, 1967, *5*, 105-212.

Lorenz, K. Z. The comparative method in studying innate behaviour patterns. In *Physiological mechanisms in animal behavior*, Symposium No. IV of the Society for Experimental Biology. London: Cambridge University Press, 1950.

Lorenz, K. *King Solomon's ring*. London: Methuen, 1952.

Peterfreund, E. Information, systems, and psychoanalysis. *Psychological Issues VII*, Monogr. 25/26. New York: International Universities Press, 1971.

Robertson, J., & Bowlby, J. Responses of young children to separation from their mothers. *Courrier de la Centre Internationale, Paris*, 1952, *2*, 131-142.

Robertson, J., & Robertson, J. Young children in brief separation: a fresh look. *Psychoanalytic Study of the Child*, 1971, *26*, 264-315.

Rowell, T. E., & Hinde, R. A. Responses of rhesus monkeys to mildly stressful situations. *Animal Behaviour*, 1963, *11*, 235-243.

Rutter, M. *Maternal deprivation reassessed*. Harmondsworth: Penguin Books, 1972.

Skeels, H. M., & Dye, H. A. A study of the effects of differential stimulation on mentally retarded children. *Journal of Psycho-Asthenics*, 1938, *44*, No. 1.

Spitz, R. A. Hospitalism: an enquiry into the genesis of psychiatric conditions in early childhood, 1. *Psychoanalytic Study of the Child*, 1945, *1*, 53.

Spitz, R. A. Anaclitic depression. *Psychoanalytic Study of the Child*, 1946, *2*, 313-342.

Tinbergen, N. *The study of instinct*. London: Oxford University Press, 1951.

Discussion of Papers by Suomi and Bowlby

MARY D. SALTER AINSWORTH

John Bowlby began his chapter (with acknowledgments to Robert Hinde) with a statement of six ways in which studies of animal behavior can contribute to a study of human behavior. Two of these pertain especially to experimental studies, two perhaps specifically to naturalistic studies, and two to studies of both types. Let me first consider experimental studies and their contributions with particular reference to Dr. Suomi's chapter, and then turn to naturalistic studies, referring both to Dr. Bowlby's paper and to some work of my own.

I have the highest regard for experimental studies of nonhuman primates and their relevance for the understanding of human behavior. Long-term experimental control of rearing conditions is neither feasible nor ethical in human research. It is extremely helpful to those of us who have been grappling with difficult problems in studying the effects of early experience on human development (including pathological development) that well-controlled laboratory research on animals should yield findings and inferences that are so similar to ours. Nothing is more likely to convince the sceptics. The research that Dr. Suomi reported in his paper—including research by Harlow, Hinde, Kaufman, Rosenblum, and Suomi himself, and their various talented associates—seems to me to represent a particularly productive and helpful produce of cross-disciplinary fertilization. Furthermore Dr. Suomi's review is presented in the context of a fine appreciation of the problems implicit in interspecies comparisons, as well as of the usefulness of the back-and-forthing that has gone on between experimental re-

MARY D. SALTER AINSWORTH ● Department of Psychology, University of Virginia, Charlottesville, Virginia.

search with nonhuman primates and the more clinical perspective of the research on human subjects.

I am not convinced, however, that what he has called "depression" in young macaques undergoing isolation or separation experiences is homologous to or even analogous with the psychiatric syndromes that are labeled "depression," for which a better label would be "depressive illness." But let me first dwell on points of convergence rather than points of divergence.

Suomi's review prompted me to recall a monograph that John Bowlby and I published in 1954 on research strategy in the study of mother–child separation. In it we enumerated the variables that seemed to influence the effect on a child of separation from his mother. These variables had been identified both from a thorough review of the human research literature on the effects of mother–child separation and from studies that Bowlby and his research team had themselves conducted. A substantial amount of inference—or let us say good judgment—was involved in the interpretation of these research findings, since in the human case the relevant variables in any one study are likely to be confounded. Therefore it was with substantial gratification that I noted that nearly all of our major points had been dealt with in Suomi's review. Let me briefly list them.

1. Length of separation has been found to be a condition affecting the responses to separation and to reunion in all species so far studied.

2. The environment during separation has been found to be important in both human and nonhuman primates. The more depriving the environment during separation, the more adverse the effects on behavior both during the separation itself and afterward in the reunion period. In the human case even the most depriving circumstances likely to be encountered in separation are not as depriving as, say, Wisconsin vertical-chamber housing. The closest parallel is when a young child is confined in a crib when in hospital. It is well known that a separation under these conditions has more adverse effects than separation in which a child is free to move about and to mingle with his peers and substitute caregivers. Furthermore, as the recent studies of James and Joyce Robertson (e.g., 1971) attest, the availability of a responsive potential substitute attachment figure is very important both in attenuating the intensity of the response to separation and in facilitating the resumption of a relatively normal relationship with previous attachment figures upon reunion.

3. Both Suomi and Harlow have focused on the role of peer relations as supplanting or even substituting for the primary infant–mother relationship. The applicability to the comparable human case is still obscure. Certainly, children who sustain separations in institutional environments show disturbance while separated despite the opportunity to interact with peers. The presence of siblings has been observed to mute separation disturbance to some extent, however. I wonder whether infant macaques separated from their mothers in the natural environment would find peers as adequate in the role of substitute attachment

figures as they do in the relatively limited and sheltered environment of the laboratory. Van Lawick-Goodall's (e.g., 1971) field studies of young chimpanzees who have lost their mothers suggest otherwise.

4. The human separation studies highlight the importance of experiences that tend to keep the attachment relationship alive despite separation—whether through parental visits (in the case of the hospitalized child), or as the Robertsons have done through talk of the parents and of home, photographs, encouraging recollections, and encouraging the expression of fantasies through doll play. There is not and probably cannot be anything comparable in the animal studies.

5. Multiplicity of separations. Hinde's finding (Hinde & Spencer-Booth, 1970) that two separations, both of short duration, have a more lasting and adverse effect on the behavior of infant monkeys, fits very well with the implications of human research. I think here of Bowlby's (1946) classic study of 44 juvenile thieves, and Bender's study (Bender & Yarnell, 1941) of children with psychopathic behavior disorders. There is also evidence that when institution-reared children are placed in foster homes, shifts of foster home exacerbate the behavioral disturbances originally attributable to institution rearing. Let me refer briefly also to my own research (Ainsworth, 1973). We have a laboratory situation that we call "the strange situation" which includes two separation episodes which together do not exceed 9 minutes of mother-absence. Not only is a one-year-old's response to the second separation and to the reunion following it more intense than to the first, but when we repeated the whole situation two weeks later we found that the second situation evoked much more anxiety than the first. It seemed to us that repeated involuntary separation experiences even of this minor sort made for increased separation anxiety. Also relevant here is the work of my ex-student, now colleague, Mary Blehar (1974) who has evidence that full-time daycare makes for child–mother relationships of anxious quality in comparison to those of home-reared children of the same age. She attributed the effect of the often repeated, long daily separations implicit in daycare.

6. How attachment figures respond to the child's behavior after reunion has much to do with how long the disturbed behavior attributable to separation lasts. In general, the human observations (here again I rely on James Robertson) suggest that if the mother accepts and responds to the increased demands for proximity and contact typically made by a child after separation, these demands tend to decrease, whereas if she resists or rejects them, perhaps fearing to spoil him, they tend to persist longer, and in some cases last indefinitely. Hinde's work on rhesus separations tends to yield evidence congruent with the human case, and the Wisconsin work certainly shows that differential effects are attributable to differing conditions in the reunion period, or at least after the cessation of especially severe periods of isolation or deprivation.

7. Finally, there is the question of preseparation relationships and their effect on responses to separation and to subsequent reunion. This is a condition
that Suomi has particularly stressed (and he refers to Hinde's work), but here we
do not have clear-cut human findings. In our monograph, however, Bowlby and I
included this as a variable that ought to be reckoned with in research on responses to separation. For that very reason Robertson and Bowlby (e.g., 1952)
in their own work sought subjects whose preseparation family relations had not
been unduly disrupted. They provided evidence that even children with very
healthy preseparation relationships are likely to respond to major separations
with significant distress and disturbance. There is other evidence in the human
research literature that children whose attachment relationships were already
anxious before separation tend to be disturbed for longer periods and more intensely after separation than do children whose preseparation relationships were
secure. But, obviously, this kind of evidence confounds preseparation and postseparation (i.e., reunion-period) variables.

In summary, it is evident that, with a few minor exceptions, there is such a
striking congruity between the findings for human and nonhuman primates in
regard to responses to separation and to subsequent reunion that it can be only
gratifying to all who have devoted time and energy to the explorations relevant
to this problem. Within this positive context, perhaps I may be allowed to
hazard a few critical comments about Dr. Suomi's report.

The first comment pertains to his study of older juveniles. He expressed surprise that even these juveniles could be profoundly disturbed by separation if
they were placed in a very depriving environment during the separation experience. This experience seems equivalent to solitary prison confinement in the
human case—an experience that is clearly traumatic. John Bowlby and I have
been pleading for 20 years that the effects of breach of bonds implicit in separation must be disentangled from the effects of deprivation—and by deprivation
we mean, specifically, deprivation of any adequate opportunity to interact with
anyone who might otherwise become a new attachment figure. Although we
should be surprised were it found that adolescents are as profoundly affected as
much younger children by separation from an attachment figure, we would not
be surprised if they, whether human or nonhuman primates, responded with
conspicuous disturbance to solitary confinement.

Second, it does not seem to me that Dr. Suomi distinguishes adequately between responses to being separated from an attachment figure and long-term disturbances of behavior, whether this occurs in the reunion period, in the case
where the young animal was returned to his original attachment figure, or
whether it occurs with reference to his response to later opportunities to form
relationships with unfamiliar conspecifics. He skips back and forth from responses to separation and responses to reunion as though he felt that it is all the
same.

Perhaps the key to the difficulty is the focus on "depression." Spitz and Wolf (1946) were among the first to draw attention to a severe form of response in infants to separation from their mothers in a depriving environment. Certainly this response resembles that of young macaques under comparable circumstances. But to call this "depression" implies that it is akin to depressive illness in adults. Instead of describing a response that occurs in a context that can be well specified, the label "depression" is noncontextual. It seems to me that John Bowlby is on much safer ground in labeling this severe response "despair" and identifying it as one phase in a sequence of responses to breach of ties with an attachment figure (Robertson & Bowlby, 1952). Furthermore, he likens the whole sequence of responses to grief and mourning in an adult (Bowlby, 1961). There is nothing essentially pathological in grief and mourning; it is a normal response to loss of an attachment figure and when normally worked through or resolved it results in readiness to form a new attachment to supplant the lost loved one. In the case of a minority of adults this normal outcome is not achieved. Bowlby points out, however, that loss of an attachment figure in early childhood, whether it is a permanent loss or a separation so long that it seems final, is more likely to set in train processes that lead to later pathology than in the case of the adult (Bowlby, 1963). He has pointed out, however, that although early bereavement seems to increase the likelihood that an individual will respond to later perhaps more minor losses with real depression, there are other outcomes in which psychopathology is implicated. In our present state of knowledge, we can point to sociopathy—or the affectionless character—as one obviously pathological outcome, but much more common is the syndrome that Bowlby has particularly stressed, namely, anxious attachment.

My final point with reference to Dr. Suomi's paper and to experimental studies in general is that they tend to assess the effects of various adverse early experiences in terms of general social behavior rather than specific bonds. My attitude here undoubtedly reflects my persuasion to an ethological viewpoint in which, from the beginning provided by Lorenz (1935), specific bonds and specific relationships are emphasized as the foundation of the superstructure of interaction with conspecifics. But I am influenced also by a clinical orientation toward the human case, which tells me that the most serious anomalies of social development are those that interfere with the establishment or maintenance of deep and lasting interpersonal relationships, that is, attachments to individuals. In this mood, I found myself very curious to know what would have happened if Suomi's monkeys who had been mother-reared had been returned to their mothers after having been housed individually or in pairs, and how the semiadult monkeys reared in family environments would have behaved had they been returned to their families.

Now let me change direction. I would like to stress the pertinence to research into human social development of naturalistic animal studies, and especially the contribution an ethological perspective has made to research with humans. I return to some of the points that Dr. Bowlby made in his introductory remarks.

First, let me touch on methodology. The human infant is a nonverbal organism. Traditional psychological and psychoanalytic techniques are inapplicable in infant research. They require either language or far more cooperation from a research subject than an infant can muster. Consequently, for far too many years the study of infancy was neglected. Direct observation of the behavior of the animal in its natural environment is the starting point of an ethological investigation of any species. It is also a good starting point for the investigation of human social development. The fundamental and yet difficult art of watching has been restored to respectability as a technique of scientific investigation. Furthermore, the ethological approach emphasizes the importance of observing behavior in context. Context here has two important implications. First, it is important to observe a given behavior in the context of the natural environment, to see in what situations it naturally occurs. And second, it is important to observe it in the context of the entire behavioral repertoire of the animal being observed. Only thus can the significance of the behavior be understood—its usual outcome and its function. Finally, ethology emphasizes the study of fine nuances of behavior rich in detail and this too I have found to be a very useful emphasis in the study of young humans and the people with whom they interact.

Second, I believe that the most important theoretical contribution offered to developmental research by ethology is the principle that species-characteristic behavior can most fruitfully be viewed in an evolutionary perspective. Dr. Bowlby (1969, 1973) has performed a great service in his insistence that humans as well as other animals may be genetically biased to behave in ways that promote individual and/or population survival in the environment to which the species was originally adapted and in which it evolved. He suggested that the vital biological function of the formation of a bond to a specific mother figure and of attachment behavior—the function that led it to be selected because of its survival value—is protection. In the environment of evolutionary adaptedness, protection from predators was the crucial issue, but even in the present environments in which man lives, attachment still has an important protective function.

For a psychologist this viewpoint opens new vistas. Most psychologists have been sold on the notion that the hallmark of the human species is its enormous modifiability, and this notion is easily extended to a belief that the human is infinitely flexible and able to adjust to any environmental circumstance. Therefore, if an infant's insistence on keeping his mother figure accessible to him proves to be inconvenient, it should be possible to train him out of it. Now, however, we are beginning to realize that it is in the mismatch between the genetic biases underlying the behavior of the infant and young child and the demands placed upon him in his environment of rearing that much of psychopathology originates. Developmental anomalies are likely to occur in proportion to the extent that the rearing environment differs from the original environment of evolutionary adaptedness, and they are especially likely to occur when the rearing environment cuts across the grain of behavioral tendencies that are deeply rooted in the species because of their important survival function.

Let me give a few examples from my own work and the work of others. The research in which I am engaged is an intensive, naturalistic, longitudinal study of the development of attachment in the first year of life. My sample consists of 26 white, middle-class infant-mother pairs. In the course of this research I have been systematically exploring the course of development of attachment behaviors as well as the development of the infant-mother attachment relationship itself, and paying attention to the extent to which early maternal behavior influences later infant behavior, or vice versa. (Our longitudinal design enables us to detect direction of effects even though our study is correlational rather than experimental.)

The first attachment behavior studied in this way was infant crying. What Silvia Bell and I found (Bell & Ainsworth, 1972) was that the amount that a baby cried in his first few months—his constitutional irritability, if you like—had no significant effect on how responsive his mother was to his crying, and indeed had no significant correlation with the amount that he cried at the end of the first year. On the other hand, how frequently and promptly a mother responded to her infant's crying during the first three months had a significant effect on how much the baby cried by the end of the first year. Early maternal unresponsiveness promoted later infant fussiness. Early maternal responsiveness promoted not only infrequent crying later, but also the development of varied, subtle, and clear modes of communication by means of facial expression, gesture and non-crying communication. These findings are contrary to assumptions that are commonly held.

I am sorry not to be able to give more detail, but I should report that multivariate analyses of a variety of behavioral measures of the infants in our sample suggests that a major dimension of the infant-mother attachment relationship is a secure versus anxious dimension, and that the groundwork for this begins in the earliest weeks of life, long before the baby can be described as attached or bonded. Furthermore, this secure-anxious dimension of infant behavior seems to be clearly related to comparable measures of maternal behavior that reflect the extent to which a mother is accessible and responsive to her infant (Stayton & Ainsworth, 1973).

Findings of this sort led me to hypothesize that the kind of environment to which a human infant's behavioral repertoire is adapted (in the evolutionary sense) includes a mother whose own reciprocal maternal behaviors are sensitively tuned to infant signals. Konner (1972), thinking along similar lines, undertook to test a similar evolutionary hypothesis. He studied mother-infant interaction in a present-day population of hunter-gatherers, who inhabit an environment similar to that in which man was believed to have evolved and who live in very much the same way that the earliest hunter-gatherers are believed to have done. This society is the San, one of the African Bushman societies. He found that San babies were carried most of the time, and in such a way that they had the same ready access to the mother's breast that the infant rhesus has. He also tells me that San babies cry very little and that their mothers tend to be very promptly responsive to their signals. By all of our traditional expectations, San children ought to have emerged as spoiled and overdependent. Not so. San 2-to-5-year-olds ranged farther from their mothers than did a comparable age group of London children studied by Blurton Jones.

Blurton-Jones (1972), also working within an evolutionary framework, posed the question as to whether the human species had evolved as a species in which the mother caches the infants in a safe place and returns from time to time to feed them, or as a carrying species in which the mother carries the infant with her wherever she goes and feeds him more or less continuously as do monkeys and apes. He compared humans with members of caching species on one hand, and with other higher primates on the other, and concluded from a number of anatomical, physiological, and behavioral indices that the human species is indeed preadapted to be a carrying species. If so, our present-day Western environment of the crib, the infant seat, the playpen, and the four-hour feeding schedule departs quite sharply from the original environment of evolutionary adaptedness. Konner's San, on the other hand, behave as though they belong to a carrying species.

Unpublished findings of my own project (Blehar, Ainsworth, & Bell, in preparation) fit the implications of Konner's and Blurton Jones's conclusions very well. It is, however, the qualitative rather than the quantitative aspects of physical contact about which I wish to report. We identified a maternal behavior that we labeled "tender, careful holding" somewhat tongue-in-cheek because of clichés about TLC. This mode of physical-contact behavior was characterized not only by a gentle muting down of the mother's normal speed and vigor of movement but also by something akin to pacing of her physical handling to the infant's tempo of response. We found that the proportion of holding that was tender and careful in nature during the first three months was significantly and positively related to the infant's later positive response to physical contact—to smiling and other behavioral manifestations of positive affect, to cuddliness, and eventually also to active attachment behavior while in physical contact. In contrast, infants' initial individual differences in response to being held did not seem to influence the mother's later positive holding. On the other hand, infants who responded positively when held by their mothers early on did seem to stimulate their mothers later to show relatively frequent display of affection while holding their babies—hugging them, kissing them, caressing them. So here we have a virtuous cycle. A mother who handles her tiny baby tenderly and carefully engenders in him a positive response to physical contact, and in turn this positive response inspires his mother to affectionate display, which undoubtedly consolidates his pleasure in contact. But do babies so handled become spoiled and overdependent and unhappy when not in contact? No. Even during the first three months of life, babies so handled protest less frequently when put down than do babies with less pleasant contact experiences. And by the end of the first year it is clearly the babies who most enjoy physical contact who are cheerful about its cessation, and then tend to turn promptly to independent exploratory play.

At the other extreme we have evidence—which I cannot report here in adequate detail—that it is through their physical contact behavior that rejecting mothers convey their rejection to their infants (Main & Ainsworth, in preparation). Infants whose mothers rebuff their bids for physical contact or who give them an uncomfortable or unpleasant time while in contact, come to have an approach—avoidance conflict about close physical contact with their mothers. Their preprogramming leads them to seek contact as other babies do, but their experience leads them to shy away from it. What emerges as a behavioral solution is ambivalence alternating with avoidance. By ambivalent behavior we mean squirming to get down when picked up, only to seek to be picked up again, or intention movements such as approach and touching the mother's foot, followed by locomotion away from her. By avoidant behavior we mean looking away, turning away, or moving away under circumstances such as maternal invitations to interaction in which most babies respond with approach and other attachment behaviors.

In an independent project on a different sample, Mary Main (1973) found that infants who were conspicuous for their avoidant behavior at 12 months of age were significantly different in a number of ways from other toddlers 10 months later. For example, they had shorter bouts of exploratory play; they were inappropriately aggressive in their play with objects; they avoided an adult playmate who attempted to engage them in interaction, and they were still avoidant of their mothers. Upon first encounter with the Tinbergens' monograph (1972) on childhood autism, Mary Main (in press) detected a marked resemblance between her avoidant toddlers and autistic children in the nature of the avoidant behaviors manifested—gaze aversion, orientation of the body away from the significant figure, and the like. If you saw her videotapes I am sure you would see the resemblance too. She does not propose that her avoidant toddlers are or will become autistic—but if one conceives of a continuum ranging from severe autism to normal behavior, they are displaced toward the autistic end of the continuum and away from "normal."

Let me conclude by acknowledging that had it not been for an ethological-evolutionary perspective we probably would not have been able to report the kind of findings which I have reported here. I might well not have had the courage to persist in naturalistic studies of mother–infant interaction which are not popular with grant-awarding panels. I would not have been especially concerned about infant behavior differentially directed toward an attachment figure. I probably would not have studied behavior in rich enough detail to have noted, for example, the subtle manifestations of avoidance. And, even had I done all of these things, without an ethological–evolutionary perspective I would not have been able to make theoretical sense of my findings.

REFERENCES

Ainsworth, M. D. S. Anxious attachment and defensive reactions in a strange situation and their relationship to behavior at home. Paper given in a symposium on "Anxious attachment and defensive reactions," at the biennial meeting of the Society for Research in Child Development, Philadelphia, 1973.

Ainsworth, M. D. S., & Bowlby, J. Research strategy in the study of mother–child separation. *Courrier, Centre International de l'Enfance*, 1954, 4.

Bell, S. M., & Ainsworth, M. D. S. Infant crying and maternal responsiveness. *Child Development*, 1972, *43*, 1171–1190.

Bender, L., & Yarnell, H. An observation nursery. *American Journal of Psychiatry*, 1941, 97, 1158.

Blehar, M. C. Anxious attachment and defensive reactions associated with daycare. *Child Development*, 1974, *45*, 683–692.

Blurton, Jones, N. Comparative aspects of mother–child contact. In N. Blurton Jones (Ed.) *Ethological studies of child behavior*. Cambridge: Cambridge University Press, 1972, pp. 305–328.

Bowlby, J. *Forty-four juvenile thieves, their characters and home life*. London: Bailliere, Tyndall, and Cox, 1946.

Bowlby, J. Processes of mourning. *International Journal of Psychoanalysis*, 1961, *42*, 317–340.

Bowlby, J. Pathological and childhood mourning. *Journal of the American Psychoanalytic Association*, 1963, *11*, 500–541.

Bowlby, J. *Attachment and loss* (Vol. 1). New York: Basic Books, 1969.

Bowlby, J. *Attachment and loss* (Vol. 2). New York: Basic Books, 1973.

In G. Servan (Ed.), Animal Models in Human Psychobiology. New York: Plenum, 1976.

Hinde, R. A., & Spencer-Booth, Y. Individual differences in the responses of rhesus monkeys to a period of separation from their mothers. *Journal of Child Psychology and Psychiatry*, 1970, *11*, 159–176.

Konner, M. J. Aspects of the developmental ethology of a foraging people. In N. Blurton Jones (Ed.), *Ethological studies of child behaviour*. Cambridge: Cambridge University Press, 1972, pp. 285–304.

Lorenz, K. Z. (1935) Der Kumpan in der Umwelt des Vogels. Tr. in C. H. Schiller (Ed.), *Instinctive behavior*. New York: International Universities Press, 1957.

Main, M. B. Exploration, play, and cognitive functioning as related to child–mother attachment. Unpublished Ph.D. dissertation, The Johns Hopkins University, 1973.

Main, M. B. Avoidance and its implications for daycare. In R. Webb (Ed.), *Social Development and Care of Children*. Baltimore: Johns Hopkins Press, in press.

Robertson, J., & Bowlby, J. Responses of young children to separation from their mothers. *Courrier, Centre International de l'Enfance*, 1952, *2*, 131–142.

Robertson, J., & Robertson, Joyce. Young children in brief separation: A fresh look. *Psychoanalytic Study of the Child*, 1971, *26*, 264–315.

Spitz, R. A., & Wolf, K. M. Anaclitic depression. *Psychoanalytic Study of the Child*, 1946, *2*, 313–342.

Stayton, D. J., & Ainsworth, M. D. S. Individual differences in infant responses to brief, everyday separations as related to other infant and maternal behaviors. *Developmental Psychology*, 1973, *9*, 226–235.

Tinbergen, E. A., & Tinbergen, N. Early childhood autism: An ethological approach. Zugleich Beiheft 10 zur *Zeitschung fur Tierpsychologie*, 1972.

Van Lawick-Goodall, J. *In the shadow of man*. Boston: Houghton Mifflin, 1971.

Prenatal and Postnatal Factors in Gender Identity

JOHN MONEY

INTRODUCTION

In what we have heard about the effects of developmental deprivation in two species, monkey and man, it has become apparent that there is remarkable cross-species parallelism. Investigatively, for ethical reasons, there are fewer options in human than in animal studies: unplanned "experiments of nature" take the place of designed experiments.

CHILD-ABUSE DWARFISM

One unplanned experiment of nature occurs as a syndrome of unremitting separation—threat rejection, deprivation, and actual injury. It is the syndrome of child abuse, also known as the battered-child syndrome. I've followed a group of abused children in whom one of the symptoms of abuse is retarded statural growth (Money, 1976; Money & Annecillo, 1976, Money & Wolff, 1974; Money, Wolff, & Annecillo, 1972; Wolff & Money, 1973). Their condition has been variously known as deprivation dwarfism, psychosocial dwarfism, and reversible hyposomatotropic dwarfism. It is pathognomic of the condition that impaired growth and growth-hormone secretion can be reversed by a change of

JOHN MONEY ● Department of Psychiatry and Behavioral Sciences, Department of Pediatrics, The Johns Hopkins University School of Medicine, Baltimore, Maryland 21205.

domicile. Along with catch-up growth in stature, there is a corresponding catch-up of intellectual growth and behavioral maturity, though some residual impairment may be permanent.

I have mentioned this syndrome not to go into detail, but simply to add to the two types of material presented by Bowlby and by Harlow and Suomi.

Reversible hyposomatotropic dwarfism came to my attention because of my overall interest in the behavioral endocrinology (or, more precisely, the endocrine behaviorology) of childhood. I am also interested in psychosexual differentiation and gender identity/role (Money, 1973).

FILIAL AND PEER BONDS AND GENDER IDENTITY

One of the things we could eventually explore is the relationship of attachment versus separation or bereavement behavior to the differentiation of gender identity. Even on a clinical–impressionistic basis, it is obvious that a conceptual bridge is needed to connect attachment/separation behavior and gender-identity differentiation. Errors or biases in gender-identity differentiation in anatomically normal children are very much related to inability to establish healthy filial and peer bonds. I may even venture the guess that a parent's own attachment or separation problems resulting in failure to establish a healthy parent-child bond may also be involved.

PARAPHILIA

A bias or skewness in gender-identity differentiation is not restricted to issues of masculinity versus femininity. There are biases within masculinity or femininity. Diagnostically they belong under the general rubric, Paraphilia—for example, fetishism, sadomasochism, exhibitionism, voyeurism, pedophilia, gerontophilia, and so on, through the entire Greek list of them. These all have their origins in early childhood. I suspect they may well be related to the kind of phenomena we've heard of from Bowlby, Harlow, and Suomi. To illustrate from a recent and, I think, fairly typical case: I was consulted by a student whose sexual life was completely beleaguered by obsessional imagery of a sadistic nature. He related it to an experience at age four. His highly disturbed mother had taken a carving knife and held the sharp edge against his penis while his older sister held him down. She threatened to "cut it off" if he ever played with it again. That is a partial and rather special form of separation threat, but it is a separation threat, nonetheless.

INTERSEX: MATCHED PAIR

Figure 1 shows a matched pair of chromosomally male intersexes or her-maphrodites—one of between 30 and 40 matched pairs I have studied longitudi-nally. They are matched in the sense that they are concordant for diagnosis, prenatal development, and hormonal puberty. That is to say, they are concor-dant for chromosomal sex, gonadal sex, fetal hormonal history, anatomical genital defect, and onset of a masculinizing puberty. They are discordant for postnatal socio-sexual history—discordant by reason of the fact that one was considered to be a girl and assigned and reared as a girl, and the other as a boy. Both were eventually habilitated surgically and hormonally in concordance with both the assigned sex and the parallel postnatal differentiation of gender identity. In adulthood they were psychosexual opposites. If you met them, you would unquestionably accept the one as a man, a husband, and an accountant, and the other as a woman, a wife, and a secretary.

The man has a relatively small penis, but it is effective in sexual intercourse. His only psychosexual imagery and practice is that of a heterosexual male. By contrast, the woman—recall her status as a chromosomal male and a gonadal male until the masculinizing testes were removed at age eleven—has the psycho-sexual life of a female. At age eleven, it was questioned as to whether or not she should change to live as a boy, despite the almost clitorine size of the hypospadiac phallus. Since she had differentiated the gender identity of a girl, it was fortunate for her that the final decision was to leave her living as a girl. The source of hormonal masculinization was cut off by removal of the un-descended gonads. Replacement estrogen induced perfect female body mor-phology. At age 19, the vestigial vaginal orifice (a urogenital sinus) was surgically deepened into a coitally functional vagina. The only remaining disability was sterility and, if children were desired, the necessity of motherhood by adoption.

This matched pair of cases exemplifies the extraordinary importance of the early postnatal and childhood years in the differentiation and establishment of gender identity and gender role. Postnatal differentiation is superimposed on whatever it is that may have happened in the prenatal period even though con-tradictory of it. Thus, there is ample opportunity for fetal reprogramming to be contradicted by the actual programming of the juvenile years.

The scope of postnatal gender-identity programming and its pathology in human beings is probably more extensive than in the other primates. None-theless, all the primates appear to be more susceptible to postnatal psychosexual programming than do the lower mammals. The difference, however is a matter of degree. The animal model and especially the primate model can therefore be profoundly useful in elucidating problems of sexual psychopathology in human beings.

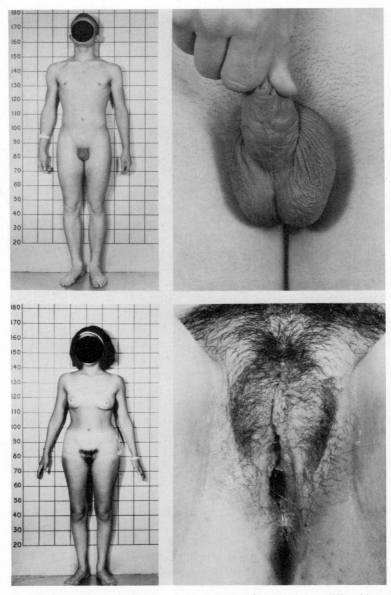

Figure 1. Matched pair of 46XY intersexes, concordant for genetic, gonadal and hormonal sex, but discordant for assigned sex, gender identity, and rehabilitative sex.

TELEOLOGY: A CRISIS OF THEORY

In my role as Discussant, let me say that I perceive a covert discontent with the principle of teleology in the preceding papers. It is not too unusual for experimentalists to eschew teleology, but less so for clinicians. Yet, I infer the discontent even in Bowlby's clinical formulations, and I totally concur.

At the present moment of history, it is almost a platitude to speak of a crisis of theory so far as human clinical psychiatry and psychology are concerned. The crisis seems to be engendered by the failure of psychoanalytic doctrine to have fulfilled its promise *in toto*, and the disillusionment thereby produced. Psychopharmacologic theory is no substitute, for it is itself still fragmentary and not comprehensive. Behavior modification generates an excessive overoptimism, simultaneously with a general feeling of discontent that it is being oversold in the same way that psychoanalytic theory was oversold in its early days.

Sometimes I think we need a new Descartes. He got us off the hook of the sacred versus the secular, theology versus science, and the mind versus the body by the fiction of the pineal gland as the meeting place of the spiritual soul and the animal body. The body belonged to science and mechanism, the mind and the soul to the church and teleology.

Descartes' solution to the mind–body problem was politically and theologically neat and expedient in his day, but it was not scientifically expedient, and still is not. The separation of mind and body enters into all of our terminology and scientific thinking. It bedevils us no end, even though we protest to the contrary.

In my own theoretical thinking, I have decided that I do not need to be hindered by the teleological concept of motivation, drive, urge, or need as a basic theoretical concept in formulating anything that I need to formulate with respect to my work with human beings, their psychodynamics, and their psychological growth and development. Of course, I'm able to listen to patients who tell me that they have a feeling, urge, or drive, and I accept their own wording. But I do not incorporate such wording as a major theoretical principle in my own overall thinking. In fact, I think of instinct, motivation, urge, need, or drive as being as useless to behavior as was the alchemical principle, phlogiston, to chemistry after Priestly discovered oxygen in 1774 and demonstrated that burning requires oxygen, not phlogiston.

I believe we can understand human beings and their behavior in much the same way as we approach the solar system (Money, 1957). We no longer invoke cosmic forces or teleological powers within each of the planets to keep them going and in place. Rather, we accept the planetary system as a going concern in itself. It is a dynamic or kinetic system, and it has continued that way ever since we were here on earth to observe it. That's true whether we look at the system as a whole, or whether we look at each planet or the sun as a separate entity.

So it is also with human behavior. I think we may assume from the outset that it is a system. It is a system of people and their interaction (in current actuality or retrieved from the schemata of memory). Like the solar system, it is by definition a kinetic or dynamic system. It is kinetic or dynamic even when one considers one specific individual within the system and examines that person as an individual, just as one might examine a single planet. Then one sees the kinetics and dynamics in interrelationships with other people because they are represented in the memory of the individual concerned.

Saul Rosenzweig, my first teacher when I arrived in the United States from New Zealand in 1947, some years ago proposed the term idiodynamic (Rosenzweig, 1951; 1958) to refer to the individual specificity of one person's behavioral interactions, past, present, and prospective. By contrast, those dynamics of interaction that are characteristic of the system as a plurality might be called sociodynamic.

Idiodynamics and sociodynamics can be studied in all species, and related to both psychologic health and pathology. The applicability of inferences from one species to another cannot be assumed, but needs to be tested.

NATIVISM AND NURTURISM

Table 1 embodies in a 2 X 2 X 2 arrangement a system designed to avoid the false dichotomies of body versus mind, of biology versus culture, and of the inborn versus the acquired. You see that in Table 1 the dichotomies of determinism are naturistic versus nurturistic (i.e., environmental), idiodynamic (or personal and biographic) versus phylodynamic (or species-shared); and imperative versus optional. The contents of Table 1 pertain to sex and gender identity, but the system can be applied also to fighting and aggression, and to separation and attachment behavior, and so on.

The system of Table 1 makes it clear, and perhaps somewhat surprising, that there are some aspects of behavior which, though nativistic, are optional. Usually one's habit of thought is to equate the nativistic with the biological, and to consider it as having the attribute of an eternal verity, fixed, imperative and without an alternative. In Table 1, however, you can see that one of the nativistic cells is a species-shared option, namely, the size of the population, the fertility rate, and the sex ratio. All three are subject to environmental influence, either fortuitously or by plan.

By contrast, it is possible to have an idiodynamic determinant, unique to an individual biography, that is not environmentally optional, but nativistically

**Table 1 Determinants of Gender Identity/Role: Examples Classified
According to a 2 × 2 × 2 Scheme**

		Nativistic	Nurturistic
Phylodynamic (species-shared)	Imperative	Menstruation, gestation, and lactation (women) vs. impregnation (men).	Social models for identification and complementation in gender-identity differentiation.
	Optional	Population size, fertility rate, and sex ratio.	Population birth/death ratio. Diminishing age of puberty.
Idiodynamic (individually unique or biographic)	Imperative	Chromosome anomalies, e.g., 45X; 47XXY vestigial uterus. Vaginal atresia.	Sex announcement and rearing as male, female, or ambiguous.
	Optional	Getting pregnant. Breast feeding. Anorexic amenorrhea. Castration.	Gender-divergent work. Gender-divergent cosmetics and grooming. Gender-divergent child care.

optional: the example is the fortuitous option that an embryo will lose or gain a chromosome. Thus one has the 45X individual with Turner's syndrome; or the 47XXY with Klinefelter's syndrome; or the 47XYY, known only as the XYY syndrome. Each syndrome is characterized by its own pattern of probable behavioral correlates (Money, 1975; Money, Annecillo, Van Orman, & Borgaonkar, 1974).

The XYY pattern has been sensationalized in the media as criminal, whereas the truth is that XYY people are impulsive and behaviorally immature for age. They may or may not get in trouble with the law.

In the nurturistic column of Table 1 you see a cell showing some determinants that, though cultural in origin, are phyletic imperatives—which goes contrary to common assumptions—such as cultural exposure to the establishment of one's own native language. Similarly, cultural example is imperative to the establishment of one's own gender identity, regardless of what the prenatal component might have been (Money & Ehrhardt, 1972).

The nurturistic column of Table 1 has an entry showing that it is possible for an idiodynamically unique determinant to be an imperative. Thus it is imperative for every baby to have a publicly assigned sex of rearing. Everyone takes sex assignment for granted until confronted with the evidence of intersexuality. Intersex matched pairs (see above) show conclusively that gender-identity differentiation is strongly dependent on assigned sex and rearing, irrespective of other criteria of sex.

MONOZYGOTIC TWINS DISCORDANT FOR REARING

The ultimate type of case for demonstrating the importance of the postnatal and social phase of gender-identity differentiation is one not of intersexuality, but of an anatomically normal baby who is sex reassigned. Such an occurrence is extremely rare and is a sequel to a circumcision or other accident that causes a total and complete ablation of the penis. Surgically, it is not possible to replace a missing penis with one that is coitally satisfactory and functional.

Two cases of ablatio penis resulting from a circumcision accident have come to my attention in which the decision was made to rehabilitate the baby as a female. In one case the child is a monozygotic twin who now, as a girl, is developmentally divergent from her brother in sexually dimorphic behavior and gender identity—she with a feminine and he with a masculine identity (see Money, 1975).

SEQUENTIAL DETERMINANTS OF GENDER-IDENTITY DIMORPHISM

Figure 2 particularizes, with respect to gender identity, the principles illustrated in Table 1. In this diagram, one can trace the components of gender identity differentiation from the chromosome complement at conception to adult gender identity. Note especially that in fetal life there is a dimorphic influence of sex hormones on external genital morphology and neural pathways. I have dubbed the principle involved the Adam principle: add something to get a male. Nature's first preference and disposition is to create Eve. The additive principle is, for the most part androgen, the male sex hormone. In behavior, the Adam principle shows itself, if present in a genetic female intersex, as tomboyism. There is no word in English for its counterpart in a genetic male, nor even a word for its deficit in a genetic male. Note also in Figure 2 that the genital appearance programs the behavior of other people and eventually becomes a visual and tactile reinforcement of their behavior to the child himself or herself.

Juvenile gender identity is differentiated, incipient errors and all, in advance of puberty. The hormones of puberty change the body and lower the threshold for the emergence of erotic imagery and practice. They do not program the type of practice, nor the sex of the partner.

The importance of imagery to human erotic function and sexual practice limits the extent to which problems of human psychosexual and gender-identity pathology can be investigated using an animal model. The primate model, however, can be used to elucidate more concerning prenatal hormonal effects, and the effects of infantile attachment and separation experiences on subsequent

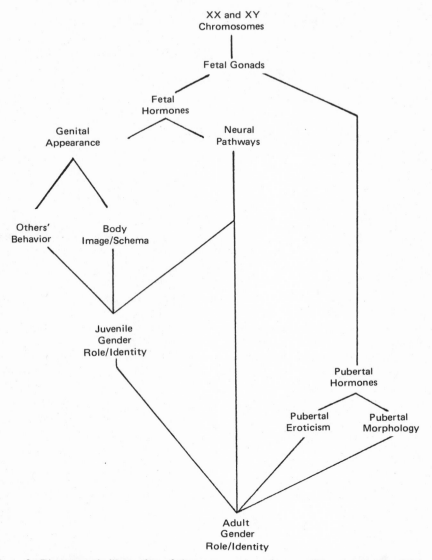

Figure 2. Diagrammatic illustration of the components and sequential programming of the differentiation of gender identity/role.

pair bonding and sexuality in adolescence and adulthood. The sexual play of infants and juveniles together can, at the present stage of social history be studied more readily in animals than man, because of the legal and moral restrictions that prevent the gathering of human data.

CONCLUSION

To maximize the utility of the animal model for the study of gender-identity and its disorders, it is imperative to have a theory applicable to both animals and man. The purpose of Table 1 and Figure 2 is to summarize such a theory, presented so that the points of cross-species similarity and difference can be readily identified.

SUMMARY

Early infantile deprivation has closely parallel effects in monkey and man. In man, deprivation effects can be developmentally devastating and may include statural dwarfism and intellectual stunting. Early deprivation and impairment of filial and/or peer pair-bonding may have adverse effects on subsequent romantic pair-bonding and sexual and erotic functions. The study of matched intersex pairs demonstrates the extraordinary degree to which gender-identity differentiation takes place postnatally, and points to the early years as the time when paraphilias may be programmed into the system. Paraphilias in human beings involve erotic imagery. There is no way, at the present time, to study imagery in animals. For an animal model to be maximally useful, therefore, it is necessary to have a theory which is applicable across species. Such a theory should be not teleological and not motivational, but it should be concerned with the kinetics and dynamics of behavior, and with regulations of the thresholds that determine whether a given program of behavior will be activated or not.

REFERENCES

Money, J. *The psychologic study of man*. Springfield: Charles C. Thomas, 1957.

Money, J., Gender role, gender identity, care gender identity: Usage and definition of terms. *Journal American Academy of Psychoanalysis*, 1973, *1*, 397–403.

Money, J., Human behavior cytogenetics: Review of psychopathology in three syndromes, 47XXY; 47XYY and 45X. *Journal of Sex Research*, 1975, *11*, 181–200.

Money, J. Ablatio penis: Normal male infant sex reassigned as a girl. *Archives of Sexual Behavior*, 1975, *4*, 65–71.

Money, J. The syndrome of abuse-dwarfism (psychosocial dwarfism or reversible hyposomatotropinism): Behavioral data, 1976.

Money, J. & Annecillo, C. IQ change following change of domicile in the syndrome of reversible hyposomatotropinism (psychosocial dwarfism): Pilot investigation, 1976.

Money, J., & Ehrhardt, A. A. *Man and woman, boy and girl: The differentiation and dimorphism of gender identity from conception to maturity.* Baltimore: Johns Hopkins University Press, 1972.

Money, J., & Wolff, G. Late puberty, retarded growth and reversible hyposomatotropinism (psychosocial dwarfism). *Adolescence,* 1974, *9,* 121–134.

Money, J., Annecillo, C., Van Orman, B., & Borgaonkar, D. S. Cytogenetics, hormones and behavior disability: Comparison of XYY and XXY syndromes. *Clinical Genetics,* 1974, *6,* 370–382.

Money, J., Wolff, G., & Annecillo, C. Pain agnosia and self-injury in the syndrome of reversible somatotropin deficiency (psychosocial dwarfism). *Journal of Autism and Childhood Schizophrenia,* 1972, 127–139.

Rosenzweig, S. Idiodynamics in personality theory with special reference to projective methods. *Psychological Review,* 1951, *58,* 213–223.

Rosenzweig, S. The place of the individual and idiodynamics in psychology: A dialogue. *Journal of Individual Psychology,* 1958, *14,* 3–20.

Wolff, G., & Money, J. Relationship between sleep and growth in patients with reversible somatotropin deficiency (psychosocial dwarfism). *Psychological Medicine,* 1973, *3,* 18–27.

WORKSHOP I: Instinctual and Environmental Learning

Edited by HARRY F. HARLOW AND STEPHEN J. SUOMI

Joseph Sidowski: I wish to comment briefly, directing my remarks to what has been written already. First, regarding sex differences: we have data which indicate that susceptibility to social stress induced during the first six months of a rhesus monkey's life results in less abnormal behavior in the female than in the male. The male reacts more to this stress over the six-month period than does the female. Second, I would like to address myself with regard to emotional susceptibility to stimulation during infancy and also to individual differences among infant primates. Anyone who has ever observed infant primates over a period of time learns to recognize and differentiate among individuals easily. This applies both to field studies, as well as within a laboratory, where one can watch an infant from birth throughout the first six months virtually daily and note that these individual animals respond differently. Great individual differences are apparent, whether the animals are males or females.

Regarding emotional susceptibility, Drs. Harlow, Suomi, and I conducted a study in which a fear stimulus was introduced at one month and presented daily for five months. It appeared that the monkeys were most susceptible to this fear stimulus at approximately three months of age. They reacted more at about three months than they did before or after that. Again, there were individual differences.

HARRY F. HARLOW and STEPHEN J. SUOMI • University of Wisconsin Primate Laboratory, 22 North Charter Street, Madison, Wisconsin 53706. (Workshop moderated by Harry F. Harlow.)

The results were interesting in that not all animals reacted the same, even though they were all the same age. Some reacted by withdrawing—they would withdraw to a corner of the cage, or they would retreat and cling to another animal. Other animals didn't respond that way. Some tried to get out of the cage. If you looked at these animals from birth, you could easily determine that some were more active than others. During a normal run of daytime events, some played more than others. Some sat in the corner more than others. So I would like to ask for consideration and recognition of individual differences. You could classify N number of animals in the laboratory into withdrawers early in response to a stress situation and classify others as more active. Then you could study these different groups individually and observe particular reactions over a period of time. You might get more specific information concerning their responses as the animals grew older. A third point I would like to make concerns some previous remarks made by Dr. Bowlby and Dr. Ainsworth. I would like to ask whether or not separation doesn't do something good. Is it all bad? Where's this old concept of frustration tolerance? I can again draw on infant monkey data. A rhesus monkey subjected to social isolation, or other types of stress, and then put in a social situation at the end of the first six months of life, for example, withdraws, huddles, and manifests all of the behaviors that might be categorized as depressed. However, let's take the same animal and put it in another situation—a dominance test where it now has been deprived of water or milk and has to get to a bottle along with five monkeys similarly deprived. They've all been pretrained as to how to get to the bottle. And now you put the six animals together. What happens to the animal who has had more stress versus the animal who has had less stress? Well, in every test that I've given, the animals that were stressed more, and who in the other social situations showed the most withdrawing kind of behavior in a playroom situation, were more aggressive in this particular situation, and, ultimately, were the most successful monkeys. They were the first ones up to the bottle, drinking all the water. Upon repeated testing, the same thing occurred. Furthermore, you could virtually rank them as to the amount of stress the animals had had as to who was going to get to that bottle first. So I would like to ask both Drs. Bowlby and Ainsworth if there is any indication from their data as to whether or not there are emotional situations in which, perhaps, the separation syndrome is less of a syndrome than it is a help.

John Bowlby: I think it all depends on what your value criteria are. I think there is no doubt that among human children who have suffered a very prolonged institutionalization, some certainly become withdrawn, and others become very aggressive. And I suppose that if you think that displaying intense aggression is a good way of behaving, then these children are better than they were before. But, no doubt the effects are varied. On the whole, there is more of a tendency for the female children subjected to these experiences to become increasingly clinging, whereas males become increasingly antisocial. Now, you might say the particular monkeys to which you referred are very good at making sure that they come first. But that is not necessarily the best sort of human being we want to produce.

Mary Salter Ainsworth: In response to the question "is separation ever a good thing?" yes, I think it is, but I think that one has to view this entirely in a developmental context. The kinds of separations that seem to be a bad thing, and do indeed seem to decrease frustration tolerance rather than increase it, are especially the involuntary ones that occur prematurely and last too long for the individual child to be able to sustain his particular level of development. The normal course of events in western society is the temporal extension of maternal separation periods. A very young infant can tolerate relatively brief absences from his mother. These periods during which he can tolerate absence become increasingly longer until finally he's able to tolerate a half day, a period at nursery school, a little later on a full day of school, then two months at summer camp, and so on. If there is a gradual introduction of separations that we think of as totally normal, a child can build up a feeling of confidence in his attachment figure whether that figure is momentarily present or absent. I think this confidence or trust in the accessibility of attachment figures is really the key to security.

In my samples, certainly the babies with secure attachment relationships to their mothers cried far less frequently when the mother left the room and in little everyday separation situations than did the infants who had insecure or anxious attachments to their mothers. Those in the latter condition seemed predisposed to separation anxiety, and I think this is what happens with what I've been calling a major separation. It is separation too long and too early, and here the child's confidence in the accessibility of his attachment figure is terribly shaken. In this case, separation is a bad thing. These are the kinds of separations that we've been talking about as pathogenic.

Irenäus Eibl-Eibesfeldt: I would like to add a piece of information concerning the Bushmen. I have studied the Central Kalihari Bushmen now for nearly five years, visiting them again and again. I have observed mother–child interaction, and often one hears of—or gets the impression that people consider—the Bushmen as growing up without any frustrations. It is true that as long as a sibling has not yet been born, there is a very strong mother–child attachment. However, as soon as a sibling is born, a most dramatic—not a gradual—cutoff occurs. The mother completely neglects the child, which is usually now about three years of age. The child seeks contact and manifests great sibling rivalry in all respects, but the mother shows a type of fed-up behavior pattern. She doesn't repel the older sibling or push the child away. She simply ignores its approaches and the many appeals for attention.

That brings me to the remark made just previously. In our culture, gradual weaning occurs. It is not so dramatic. The child is adapted early to gradually becoming independent. Pathologically, what happens to the child is nearly the same. There is a very strong protective phase, for three years perhaps, and then with the Bushmen a very dramatic cutoff of the ties. In our culture, you get a gradual weaning.

Robert Hinde: Although it will sound like a truism as a result of the discussion that has gone on before, I believe the following needs to be added: It is that we must never forget that development involves a constant ongoing *interaction* between the organism and its environment. When Mary Ainsworth says that the infant behaves in one way because of its prewiring and in another way because of its experience, she knows what she means. She doesn't really mean that there are two mutually exclusive elements in the infant that are causing it to behave in different ways. When I hear that imprinting produces a long-lasting trace, I begin to get a little worried and immediately want to ask: How much of the effects of this experience on behavior at time T and at time $T + N$ are due to what happens at time $T + 1, 2, 3$ in between, rather than there being a long-term trace which is formed now and has its results later. When Dr. Levine mentions male–female differences in response to the stress of isolation–separation, I begin to get rather more worried, for this reason. In analysis of our own data on rhesus mother–infant relations, we find that male infants are on their mothers more than females, up to about 10 weeks of age. However, with older infants, subsequent to about 30 weeks, the males are on their mothers less than females are. In between there is a transitional period. Therefore, one would predict that separation experiences would affect males more than females at one age, and females more than males at another age.

Finally, I want to end with a question on this same theme, which I also think relates to the basic animal-models topic. I would like to ask Dr. Money where, in terms of his nativistic versus cultural dichotomy, noncultural experience comes in. My Wisconsin colleagues are probably right about this, but as I understand Dr. Goy's data, the tomboyish behavior of androgenized female rhesus infants is dependent on social experience. Thus, there is an effect of experience that cannot be classed as a nativistic determinant or a cultural determinant. It is apparently due to males finding tomboyish behavior more reinforcing than females find it.

John Money: I am not sure if I should answer for Goy's material on tomboyish monkeys, but I'll answer in terms of how I have always interpreted it. So far as I know, the tomboyish behavior occurred in genetic females that were androgenized *in utero* as a result of their mothers being heavily injected with androgen during the appropriate sensitive or critical period of fetal development. Insofar as I understand it, the tomboyish behavior of the hermaphroditic female monkeys appeared without any special manipulation in the course of their growing up. They were raised the same way as other females and other males in the colony, and showed this tomboyish behavior in relationship to the other girl monkeys and boy monkeys. I think the issue of how much of it was nativistic and how much of it was elicited by the play with the other infant monkeys in their peer group is really at the heart and core of the matter.

In this instance it seems to me reasonably clear that the hermaphroditic monkeys as a group scored much higher on rough and tumble play, for example, or threat play, and this was correlated with the prenatal hormonal history because there was nothing else to correlate it with. This turned out to be a very high degree of probability, indeed. That certainly is true with regard to the large group of girls I've studied—girls with the adreno-genital syndrome which virilized them *in utero* to the point of giving them an extremely large clitoris and in a few instances, a clitoris that had the skin wrapped around it for a penis. These youngsters are tomboyish to an extraordinarily high degree. I think here the nativistic factor is that this peculiar hormonal event happened to them, and then, in the ordinary course of growing up, they showed their tomboyish behavior without any special control or regulation of other variables.

Robert Hinde: As I understand the Wisconsin data, the point is that the difference between males and females emerges only gradually in the course of social experience. If that social experience is delayed for a month or two, then the emergence of the male–female difference in tomboyish behavior occurs later. The clarification I wish to make, and I think it is important to the theme of this conference, is that with Dr. Money's very interesting scheme, which he discussed earlier, involving the two-by-two-by-two paradigm, he divided some of the determinants into nativistic and some into cultural. It seems to me that if you mean cultural, in the sense that an anthropologist would use the term cultural, then you are leaving out an area of determinants, namely those of noncultural experience. It was determinants of that sort which were causing the gradual emergence of the male–female difference in these rhesus monkeys.

John Money: I will answer that by saying that if there is a better word than cultural that will easily encompass all of these other determinants, I will be delighted to use it. When I made this scheme some time ago, I was specifically addressing myself to a matter of cultural versus noncultural issues. But, I am not addicted to the word, by any means.

Mary Salter Ainsworth: This is rather like a fugue. I would like to reply to Dr. Eibl-Eibesfeldt. I also observed among the Ganda, with whom I spent two years, the response to weaning that Dr. Eibl-Eibesfeldt reported. If this abrupt shift in maternal behavior contingent upon the birth of a sibling occurs at three years of age, I think it is very likely that the child will be able to take it in stride if he or she is sufficiently well developed. I am not denying that it may be upsetting to him for a while.

Among the Ganda, however, although weaning traditionally took place at about three years of age, with the influence of culturalization, the spacing of children broke down. Today it is very likely that the sibling will be born when the baby is a year old. Under these circumstances, weaning is a very traumatic thing. The Ganda babies I observed after weaning, provided they had had this kind of very close relationship with their mothers in which feeding seemed to get built into the attachment relationship, responded to weaning precisely the way babies respond to separation, with greatly increased attachment behavior and everything that would make us describe it as anxious attachment. Weaning had, in fact, made the relationship very insecure. My point is that, in terms of any objective criteria, an experience that takes place at age three is an entirely different thing than the same experience when it takes place at age one.

Seymour Levine: I am going to continue the fugue. I'm concerned about Robert Hinde's concern. I don't quite understand it, because clearly what I am saying is related to the whole process of individual differences, which is really the nature of how an organism responds to an environment. It's terribly idiosyncratic in most of the species we've been talking about. Now the male–female thing is in part one potential variable which indeed may make some difference in the nature of how we look at an individual's responses. Unfortunately, when we look at most of the studies, at least most of the nonprimate studies, in the literature, they just don't run females. Most of the time they simply ignore them. The malnutrition study is very interesting in this regard. They run males and females, don't find any differences in females, and then say they're not going to run them anymore. They do not address themselves to the nature of why this is a particularly interesting variable. To get back to Robert Hinde's point, if we take the concept of a sensitive period, we have in this data males and females showing differential times on the mother; the separation may have different effects at different times of separation, but it doesn't hold that given a sensitive period, the separation of the male and female at the same time is going to have the same effect.

Robert Hinde: My concern is over the generalization which you made. I've now forgotten the exact words in the course of the fugue, but you said something to the effect that "males are more vulnerable than females," and what bothers me is that you don't have to tell us that. We've been trying to tease apart the factors determining individual differences, both in the behavior of our infants and in their responses to separation. What we find is that the differences between males and females varies with, for instance, the dominant status of their mothers in the group and also with the age of the infants.

Seymour Levine: While we are discussing concerns, let me talk about another equally broad issue—the glib use of the term "evolutionary theory." I think we tend to be enamored of kinds of evolutionary theory, and yet in part, it's very easy to play adaptive games, to take any behavior and find an adaptive significance for it. People who advocated Freudian theory did this very cleverly for many years. Now the same thing, I'm afraid, is beginning to occur with regard to the glib use of evolution. Today we tend to use the term "adaptation" if a behavior works. You see we decry teleology, and yet we come back to the fact that if the behavior exists, we can find an adaptive role for it. I think this can lead us to a great deal of very dangerous thinking. In fact, there are some instances in nature where you can see animals killing themselves by adaption. In the act of adapting to the particulars of one system, they destroy the total system, and such instances do exist.

Sam Irwin: I wonder if any of you can relate these various phenomena: (1) to hyperactive children and attachment and separation anxiety in relation to its possible etiology, and (2) to autism that we see in children generally labeled autistic children, or autistic schizophrenia. What are the correlates here? Interestingly, among these children avoidance behavior is a primary characteristic. One wonders if this intrinsically develops in the child, or whether it unfolds from the interaction with the mother. A feature of this unfolding is that the individual needs exquisite control over its environment, even with toys that may be available. The autistic child prefers toys that it can manipulate but which do not respond to or reciprocate its manipulation.

Mary Salter Ainsworth: I mentioned the work of Mary Main, who was particularly interested in avoidant behavior in toddlers. Upon first encounter with the Tinbergen monograph on childhood autism, Mary Main detected a marked resemblance between her avoidant toddlers and the autistic children photographed by the Tinbergens, particularly in the nature of the avoidant behaviors mentioned and manifested, including the gaze aversion and the orientation of the whole body away from the significant figure. She doesn't propose that her avoidant toddlers are autistic or that they will become autistic, but if one conceives of a continuum ranging from severe autism on one end to normal behavior on the other, I think we could say that both my avoidant one-year-olds and her avoidant toddlers are displaced from the normal end of the scale somewhat toward the autistic end of the scale.

Sam Irwin: I would like to suggest to some of you who may have the opportunity for experimentation along these lines that it is possible to significantly influence the degree of involvement or attachment with distress state with drugs or with environmental stimuli. For example, in studies I've done, particularly with narcotic analgesics, I've had an opportunity to detach an individual from his own distress state to a degree you cannot achieve in any other way. In the case of autistic children, I would recommend this as very useful therapy for initial treatment to get them over this particular difficulty of avoidance of any kind of contactual behavior through this detachment and gradually lead them away from it. On the other hand, I would suggest exploration of the amphetamines, which we see are very effective in hyperactive children. In studies of 12 primate species, varying very much in activity level, we found in giving a constant dose of 0.1 mg/Kg of D-amphetamine orally, that those cases with very low levels of spontaneous activity exhibited an increase in activity; those with high levels of activity exhibited a decrease; and those in the middle exhibited no change. With that one dose of amphetamine, we were able to obtain in all species just about the same level of locomotor activity. But there was one additional observation that was particularly important: in all animals that had exhibited stereotyped or bizarre behavior, those behaviors were abolished by this very small dose of D-amphetamine which is equivalent to a human dose.

Now, the interesting thing here is that with the narcotics you can achieve a state of detachment. With a tricyclic antidepressant, in our human studies, I also observe an increased sense of detachment from one's distress state. Furthermore, one can achieve this to a greater degree with a narcotic drug. I wonder if the D-amphetamine may not actually produce an increased sense of confidence, so that individuals can then carry on behaviors they ordinarily would be overwhelmed by or fearful about. We see this very often. Therefore, one of the mechanisms by which one might rehabilitate or treat, hyperactive children with a drug like an amphetamine is to increase this level of subject confidence. An alternative is that much of the bizarre stereotyped behaviors one observes in monkeys isolated in separate cages may be the result of insufficient environmental stimulation, and so they tend to exhibit a body of behaviors in order to maintain some level for an optimal feeling state—a subjective feeling state. We know that the amphetamine in very low doses can lower the threshold for response to environmental stimulation, and so one then shows less hyperactivity. It is possible that the phenomenon works by both mechanisms. But I think in these situations, with animals or humans, involving attachments and separation and anxiety, these two classes of drugs—narcotic analgesics and amphetamine types—would be very worthwhile exploring experimentally.

Arthur Kling: I would like to make two comments particularly appropriate to the question of the relevance of the animal model to human psychopathology. One is that we recently had an opportunity to observe the behavior of six adult female rhesus monkeys who had been reared in social isolation for a period of six years at the National Cancer Institute. They were reared not in sensory deprivation but in social isolation in a relatively germ-free environment from the beginning of life. We were asked to look at the extent to which these animals had the capacity to resocialize with one another, and by what process this might occur. Therefore we put them all together one day. For the first time in their lives, they experienced contact with another monkey. The result was hyper-aggressive behavior of the type that has been reported for deprived adults by the Wisconsin laboratory. We then decided to keep the animals together for a long period of time and make detailed observations. An animal was only removed when there was a danger of it being severely wounded, and then it was removed for a short period of time, repaired, and put back into the social group. The interesting thing was that the hyperaggression syndrome and lack of appropriate or positive social interactions began to change rapidly, so that over a period of two months the aggression went down to practically nothing, while the positive social behaviors rose to levels which were not approximate, but getting close to what we'd see in an all-female group of rhesus monkeys.

When we look at human pathology in hospital situations, at people who have had a long life of asocial or disturbed emotional behavior, we do not see this kind of restitution of social interaction. What we do see is the tendency of patients to remain isolates. I'm particularly thinking of the schizophrenic syndromes. Yet, in the monkey model we see a remarkable restitution of positive social behaviors, with a decrease in aggression. After keeping these monkeys together for three months, we then separated them again, placed them in individual cages for four or five months, and subsequently reconstituted the group. While there was a very brief flurry of aggressive behavior, it quickly diminished, and the positive social behaviors came back very, very quickly to first test levels. I wonder if Drs. Suomi, Harlow, or Sidowski would care to comment on their experiences with the resocialization process of adults and the relevance for human studies? Secondly, regarding Dr. Irwin's comment about amphetamines, Dr. Miller in our laboratory has been looking at amphetamines in a rhesus social group. He has found a remarkable difference in effects in brain damaged, particularly frontal-lobe damaged, versus normal monkeys. Brain damaged or frontally lesioned monkeys show severe hyperactivity in response to amphetamines. But more importantly, the amphetamines seem to produce asocial animals. They no longer interact with each other. Have any studies been done, in terms of social interactions, with the children who have been on long-term amphetamine treatment?

Sam Corson: Changing the subject, I do want to support the contention of the behavior school of Wisconsin in relation to the possibility of what would appear to be almost complete rehabilitation of nonadaptive psychopathology, even in adults. We've had experiences with adult hyperkinetic dogs, as well as with adult dogs showing what appeared to be completely incorrigible violent behavior, where we were able to eliminate these types of behavior. For example, we treated a violent dog with amphetamines and psychosocial therapy, and after two years without drugs, the rehabilitation remains. The dog appears to be permanently nonviolent.

Harry Harlow: Sounds like a nice dog.

Joseph Sidowski: I would like to respond to Dr. Kling's statements about social recovery in isolate monkeys. I'm surprised! Over a period of two months you've reconstructed what basically is normal social behavior in six-year isolate monkeys. Is that correct?

Arthur Kling: They aren't totally normal by any means.

Joseph Sidowski: Well, it sounds rather miraculous that a period of eight weeks of social learning could produce these interactions after a six-year period of social isolation. Your response measures are probably important too, I suppose.

Stephen Suomi: Dr. Kling, I assume these monkeys were all females, because if males were involved, it would seem exceedingly unlikely that, first, aggression would not have been maintained for a longer period of time, and second, that the wounds resulting from the aggression would have healed rapidly enough for rapid reintroduction into the group. With regard to the notion of approaching normal social behavior, I think, as Dr. Sidowski pointed out, it depends on one's behavioral measures. What behavior is one looking at in this sort of situation? In the 1950s and 1960s, taking Harry Harlow's data, some attempts were made at just this sort of rehabilitation with older female isolates. The attempts took place not only at the Wisconsin Primate Laboratory, but also at the Vilas Park Zoo monkey island at the same time. Aggression among these animals did diminish over time, and in fact, some positive contact behaviors emerged. On the other hand, Dr. Harlow introduced the colony's most effective breeding male to this group to see if normal sexual activity would result. Normal sexual activity did not emerge from this particular living situation. So, again, one has to be very careful about the sort of behavioral measures that are being examined when one talks about improvement or lack of improvement, about rehabilitation or curing, as opposed to the syndrome being maintained.

Charles Stroebel: I would like to direct a question to Drs. Suomi and Harlow. Cardinal clinical features of depression in the human are abnormalities in the sleep–rest cycle, rest–activity cycle, clinical insomnia, and appetite disturbances. I'm curious as to what extent you have observed these in your monkeys, and I think it would be important to draw that parallel today, in view of the comparison we're trying to make between animal and human models.

Stephen Suomi: Up to this point, we have not made normative observations on either sleeping patterns or amount of food intake, or in general, eating patterns, following these so-called depression-inducing manipulations. However, our casual observations are as follows.

First, animals separated from mothers or peers look as if they haven't been sleeping well. Their eyes are red. They appear inactive. To what extent this is a function of the separation itself, and to what extent it's a function of lack of sleep, if in fact it occurs, we're not sure. Again, up to this point we have not undertaken formal examination of sleep. Second, with respect to food intake, we feed our monkey subjects once a day, late in the afternoon. They receive a specific amount of monkey chow, and for most normal animals this chow is gone from the cage within several hours. Among the animals that have been separated, however, or put through other so-called depression-inducing manipulations, we find that half the food is still in the cage the next morning. That is, over a 24-hour period, the total amount of food consumed may not differ from normal subjects to these depressed animals, but certainly, the rate at which they eat seems to be affected, and we are currently planning to take somewhat more formalized measures of eating activity in the future. I don't know how we're going to do this, but it certainly merits consideration.

Charles Stroebel: Could you comment then on the physical health of these animals, i.e., appearance of symptoms which you might call psychosomatic— funny breathing patterns, neurogenic skin lesions, gastrointestinal distress, that sort of thing?

Harry Harlow: Certainly you do not get that kind of symptom like you do in human beings. We had one animal that went through prolonged sleeplessness for some periods of days. This is an exception. And we had several animals that were isolated for 90 days from birth. Upon removal from isolation, they died of self-induced anorexia. But that's not the rule. To prevent future occurrences, we implemented the simple technique of force-feeding.

Joseph Sidowski: I want to make one point. It's not that easy to define the sleep pattern without certain kinds of psychophysiological measurements, especially if you look at animals in isolation, for example. They tend to huddle and self-clasp, and often it's difficult to tell when they're asleep and when they're awake. So again, I think it's a matter of finding that response measure that provides an adequate and scalable measure of sleep.

It is interesting to note that subhuman primate EEG patterns during sleep are quite similar to man's. In contrast, certain features of cat paradoxical sleep are modified, or absent in humans. After short adaptation, macaques sleep eight hours a day and display approximately 20% paradoxical sleep. As you know, paradoxical sleep refers to a period during which EEG measures indicate an active brain, even though the animal is not awake. The waking brain and deep sleeping body discrepancy defines this sleep phase, also characterized by REM sleep. It is surprising to find this similarity in the macaque and man, since the macaque is an aggressive omnivore that is far less peaceful and tractable than most laboratory animals. On the other hand, the chimpanzee, the omnivore most similar to man, is not known for its viciousness, although it is a formidable fighter under certain circumstances. The laboratory chimp, after short adaptation, is reported to sleep 11 hours a day, three hours longer on the average than man, and show 19% paradoxical sleep. The baboon, in many ways similar to the macaque behaviorally, is a strong and aggressive fighter that also exhibits the paradoxical sleep phase. However, the proportion of total sleep time indicating paradoxical-type activity is lower than that of the macaque and man. In other ways baboon sleep patterns appear similar to "poor sleeper" patterns in humans.

It appears that REM deprivation may contribute to depression and other psychological disturbances, although the relationship between sleep disturbance or REM interference and psychopathology is not yet clear.

Lawrence Kolb: Dr. Money's unusual twin case brought to my mind the study of twins that my colleagues and I made some years ago. I think this twin study brings up one area in which the relationship between the human and the animal studies points up a serious difficulty in our ability to resolve or weigh the significant factors. We took off from probably one of the worst twin studies that Franz Kallmann ever conducted—the study that dealt with homosexuality in twins. My colleagues and I decided that we would attempt to collect a group of identical twins that were divergent for homosexuality. And, indeed, we found five throughout the United States. We saw the twins themselves, their parents, and a number of friends. Now the interesting factors in this divergent twin development were these. First, the mothers gave birth to a set of twins of the sex opposite that which they desired. Second, the mother identified one of those twins as in some way physically weak and more desirable. Third, she spent most of her time with this particular twin. Fourth, at adolescence, she also prevented this twin from developing associations with the opposite sex. I might point out that in the majority of instances, the mother named this twin according to the sex desired in her own fantasies. In other words, there was an identification on her part of one member of that twin pair to grow up in the sexual role model that she felt desirable. One further bit of information about the eventual development was this. The twin who had the greatest attachment to the mother and vice versa in adult life seems to have adapted better socially than the other twin, who grew according to the genetic gender determination. What I'm asking here is, how do we weigh these various factors in terms of eventual human pathological development? Furthermore, I'd like to ask Dr. Money one question. I too am a bit puzzled by the split between nativistic and cultural. Does he include familial under cultural?

John Money: Yes, I guess I would, unless we're talking about familial in the sense of genetic strain.

Jose Delgado: In the discussion we're having this afternoon about the possible relevance of animal models in human psychopathology, there are two elements which should not be omitted from the discussion. One is that we are dealing with monkeys as if they were some kind of homogeneous groups, which they are not. Monkeys are not all the same. I think we should consider studying the individual reactions of these animals. In our studies we try to run an individual profile to determine how individual monkeys are alike or different from each other. I think this is a very important element, which usually is left out of papers, and sometimes symposiums.

There is a second element which I think is also very important in terms of evaluating reactivity and understanding our results, and that is the social position of the animal within the group. In other words, when we give drugs, or when we study behavior, it is very different if we are dealing with the boss animal or with the lowest animal in the group. In order to try to understand and to evaluate the different reactions to drugs according to the hierarchical position, we manipulate the hierarchical structure of the colony and study the boss of the group in the same manner as one low in the hierarchy. I think these two elements—individuality and social-hierarchical position in the group—should be considered, and I would like to know the feelings of the panel about these two elements.

Harry Harlow: That problem has not been overlooked. We have an enormous amount of data collected over a 40-year period. The social measures are interesting. There is no good measure of dominance. There are five or six or seven different measures, but none of them correlates with any other. It is a very tricky thing to do.

Jose Delgado: You are right, as usual. But even so, I think at least we could give some indication of results. First, priority in obtaining food. This is not a total proof of dominance, but it is at least a situation in which you can evaluate the social position of the animal. Second, the number of threats, attacks, and their direction. In some of our experiments we have been able to determine that the same animal, according to his own hierarchical position, manifests differing reactivity to brain stimulation.

Robert Hinde: While I am in absolute sympathy with what Dr. Delgado wants to do, I will independently echo our Chairman's strictures about dominance. Recent field studies of macaques have shown that dominance is only one, and a relatively unimportant, element or aspect of social structure. Kinship relations are infinitely more important in determining the social interactions of field-living primates than is dominance status. The task in understanding social structure seems to me to be to find a means of integrating the effects of the three categories we now know to be important—age and sex status, dominance status, and kinship relations—and see how these together determine the affiliative relations between the animals one observes. We must not be trapped by this dominance thing; we've been trapped by it for too long.

Seymour Levine: There are all kinds of pleas which we can make. Every time I walk through a zoo I get horribly depressed because I see animals in cages that are totally abnormal animals. And yet every study we're talking about is being done on an animal that lives in a cage, and this is totally abnormal. In a sense, the kinds of questions we are asking are analytical questions which are very difficult under any other circumstances, but we want to take certain factors into account. We are already dealing with an almost pathological organism to start with.

Harry Harlow: We know what makes the difference between a normal and an abnormal monkey–not a depressed monkey, just an ordinary caged monkey. Cages aren't very pretty, and the animals reared alone in them aren't very pretty either. If you do research you have to choose what measures you're going to take; that's all there is to it. You pray that they have validity; you don't know whether or not they have reliability, but you just have to gamble!

Jules Masserman: After Jose Delgado came all the way from Spain to be with us, somebody ought to support him. There's not only a tremendous difference in individual reaction to stress, there are species differences that are most dramatic. Subjected to exactly the same sort of conflict or uncertainty in experimental situations, a rhesus monkey will develop various behavioral aberrations and fall from social grace and so on; a spider monkey will regress to the point of nestling back into the arms of the experimenter, where no adult self-respecting spider monkey would think of being; or a vervet monkey will become homicidal. Now within the various species, tremendous individual differences can be observed that can be correlated only, I suppose, by extrapolation to genetic and experiential difficulties and differences in the individual animals. Thus, with the matrix of variation, it takes a very great deal of experience and many animals, to tease out the vectors concerned.

Conflict of Adaptation to Changed or Induced Environmental Conditions

Phylogenetic and Cultural Adaptations in Human Behavior

IRENÄUS EIBL-EIBESFELDT

THE COMPARATIVE APPROACH

One method in ethology is the comparison of animal and human behavior and it is thus often accused of jumping to conclusions about man from the observations of animals.

When the male flightless cormorant of the Galapagos Islands returns to his mate after fishing to take his turn in brooding the egg or protecting the young, he brings along a present, a bunch of seaweed or a starfish, for example. He holds his gift dangling from his beak and approaches slowly, passing it over to the female, which grasps the token and inserts it, with nestbuilding movements, into the nestmound. After this is done she allows her mate to dry his wings nearby and finally to take over the nest.

I repeatedly took advantage of the fact that this island race of birds lacks any fear of predators, and took the present out of the beak of the approaching bird before he had a chance to pass it over. This puzzled the bird for a moment, but he would continue onward to his mate. Once he arrived there, a little drama occurred. His mate would attack the oncomer who had no present and regularly drove him away. He would then search for a piece of wood and only then approach anew. Coming with the present he was regularly accepted. Evidently, the

IRENÄUS EIBL-EIBESFELDT • Arbeitsgruppe für Humanethologie, Max Planck Institut für Verhaltensphysiologie, D-8136 Percha Kreis Starnberg, West Germany.

passing over of nestbuilding material was an important part of a greeting ceremony, serving the function of appeasement (Eibl-Eibesfeldt, 1965).

If two chimpanzees meet each other out in the wilderness on friendly terms,
they may embrace each other and press their lips upon each other in a fashion
resembling a kiss. If a lower ranking chimp approaches a higher ranking one he
extends his hand with the palm facing upward and he may show a facial expression resembling our grin. The high-ranking chimp normally accepts the approach
by touching the extended hand of the other one in a greeting gesture of reassur-
ance (van Lawick-Goodall, 1968).

The behavior patterns just described show striking similarities to certain patterns of human greeting behavior. They indeed share a number of features. How
are these similarities to be interpreted? Can we learn from the study of animal
behavior anything relevant for the understanding of human behavior?

Certainly, the similarities are puzzling and need to be explained. In most
cases they are not just accidental. Comparative morphology has shown that
similarities can either come about as an analogy by similar selection pressures
shaping a structure in different species independently along the same line. The
parallel adaptation of body form in fish, marine reptiles, penguins, and whales is
a well-known example. Often, however, the structural similarity is inherited
from a common ancestor that already had the structure observed in the two
compared species. Thus the vertebrate column is a structural element which all
vertebrates inherited from a common ancestor. The methodology by which
homologies are discovered has been worked out in detail by morphologists.

The three main criteria are the *criterion of position*, the *criterion of specific
quality*, and the *criterion of linkage by intermediate forms*. Cranial bones, for
example, vary considerably in their form; however, the position within the other
bones allows us to identify a nasal or parietal bone. The criterion of specific
quality deals with formal similarities, which often do not suffice to indicate
homology when taken alone, since similar patterns can derive from a similar
selection pressure shaping the pattern in different species along similar lines. By
probability, however, this can be excluded in some cases, e.g., in more complex
expression movements.

The most valuable criterion of homology is the linkage by intermediate
forms which permits us to homologize even quite dissimilar patterns. The linkage can be discovered by studying fossils or related recent species. Since behavior
patterns in many ways are similar to morphological structures, the criteria of
homology can be applied for their study (Wickler, 1967).

It is generally accepted that the study of homologies may contribute to an
understanding of man. The value of primate studies has been repeatedly emphasized. Studies of geese and cichlid fishes, on the other hand, are often thought to
be of little relevance to the understanding of human behavior. Geese developed
their complicated social behavior in a parallel, independent evolution to man.
Similarities are "just" analogies.

Those who argue this way miss the all-important fact that by the study of analogies we learn about the laws which, independent of the phylogenetic relationship of the compared species, govern the development of a particular structure. It is legitimate to be interested in the laws that govern the construction of wings. The biotechnicians study wings of as many different species as possible. It does not matter whether these are wings of insects made from cuticular folds, or wings of vertebrates modeled from a pair of extremities. These structures can even be compared with man-made wings of aeroplanes. The basic laws of aerodynamics govern these different flying organs and it does not even matter whether they are culturally or phylogenetically evolved.

The same, of course, holds true for behavior patterns. If one is interested in learning about the laws determined by function, which gave origin to a particular social structure or phenomenon—monogamy for example—one would be ill-advised indeed to study our closest relatives, which simply did not develop this type of pair bond. However, it can be useful to study monogamy in as diverse species as can be found and should it be discovered that monogamy in all the different species is following the same laws of function, then we may derive working hypotheses that help us understand the phenomenon in man (Wickler, 1972). Of course, whether these working hypotheses indeed hold true for man has to be found out by the study of man.

Returning to our examples, we will find, by applying the criteria of homology, that the chimpanzee is in the possession of behavior patterns homologous to man. Kissing is in both species a behavior pattern with a signaling function derived from the pattern of mouth-to-mouth feeding.

The greeting ritual of the flightless cormorant, on the other hand, is independently developed. In both cases, however, the function of appeasement and the means by which this is achieved are fairly alike. Indeed, if we compared a greater number of greeting rituals in many more species, we would find out that there are not so many different ways by which appeasement can be achieved. Most commonly, patterns derived from the mother–child interaction repertoire are used by adults to appease and finally bond, not merely within the vertebrates, but even in insects. And if we compare animal rituals with cultural rituals of man, we may find that they are shaped principally along the same line, a fact that should be paid more attention.

To provide just another example: Tinbergen (1959) discussed a number of appeasement postures used during courting and greeting in birds which are based upon the principle that peaceful intention is expressed by demonstratively turning away a weapon or presenting it in a posture where it can do no harm. Boobies, for example, point with their beak toward the sky during courtship. Another example of the turning away of aggression-provoking signals is done by the blackheaded gulls who, during the initial stages of courtship, face away and thus avoid the black face mask, which is a threat-releasing signal when fully shown to the opponent.

And indeed, many a cultural pattern of greeting follows this principle. Presenting arms at a state visit, for example, clearly demonstrates peaceful intent. The gun is brought into a position where it could not be a threat. A Masai during greeting thrusts his spear into the ground in front of him, again following the same principle. A soldier visiting a house in our culture is expected to take off his pistol belt.

PHYLOGENETIC AND CULTURAL ADAPTATIONS

Phylogenetic Adaptations

By definition, a structure is "adapted" if it contributes to the survival of the species. They thus deal with problems forced upon the organisms by the environment and depict certain aspects or features of the outer world. The hoof of the hourse, as Lorenz once said, depicts the steppe; that is to say it is modeled as a locomotor organ to fit this special type of ground. In basically similar ways social behavior patterns fit features of the social partner. Birdsongs, for example, fit the receptor devices of the conspecific tuned for them.

If we do not simply assume that patterns were God-given, we have to ask the question how adaptation came about. Logically, we must assume that at some time, by a process of interaction between the environment and the organism, information concerning those features toward which adaptation took place was acquired by the organism. Two ways are known by which organisms can acquire and store information. One is the process of individual learning either by direct interaction with the environmental features in question or via a mediator (cultural tradition). The other process of information acquisition takes place during phylogeny by the well-known processes of mutation, recombination, and selection. In the latter case, the information is stored in the genome of the species and decoded during ontogeny.

Whether adaptation on a well-defined level is due to phylogenetic or ontogenetic acquisition of information can be checked by means of a deprivation experiment (Lorenz, 1961).

When we hear two birds singing an identical song, we can ask which common source of information both have tapped. If we deprive them during ontogeny of hearing songs of conspecifics and if they nonetheless produce the species-specific stanzas and song phrases, then it is proved that the information concerning this patterning was encoded in the genome. No further information concerning the patterning needs to be fed into the organism during ontogeny. Although the process of development is still to be studied, the basic question whether it is a preprogrammed event—a process of self-differentiation—or not is answered.

The concept of phylogenetic adaptations replaced the old concept of instinct, although the terms instinctive and innate are still used as a convenient shorthand description synonymous with phylogenetically adapted.

The concept has been challenged until recently. Those who think I am overstating the case may read Skinner's (1971) recent contribution or, as another example, the discussion of Hailman (1967). Lorenz's *Evolution and Modification of Behavior* (1965) was thought to be simply digging out the old nature–nurture controversy and thus has been considered as a step backward. The foregoing paragraph made it clear, I hope, that this is not in fact the case, but that we have to consider this contribution as a thorough reformulation of the problem, admittedly inspired by Lehrman's (1953) most valuable critique.

Since Lorenz (1961, 1963) clarified the ethological point of view, no one has disproved the core of the argument on logical or experimental grounds. Attempts, however, to sidestep the argument by shifting emphasis and thus blurring the issue are numerous. The argument used goes like this: Well, it's not so important to know that after all; it is only important to study ontogeny. There is always an interaction of innate and acquired components, of course, but both intermingle in a diffuse way, and there are therefore no techniques to separate them. Some even call themselves interactionists (Hailman, 1967), but apparently the meaning does not seem very clear to them and by their emphasis on the impossibility of analyzing the contribution of the genetic and individually acquired components they adopt the position of defeatists of research.

Charlesworth (1975) has very aptly criticized the unclear thinking of representatives of this new trend. He writes:

> Being an interactionist, though, who denies the utility and reality of the innate–learned distinction may not be saying much about how critically one has examined the nature–nurture issue or how fruitful one's research may be because he vigorously asserts interactionism. Those interactionists who write off the nature–nurture issue as dead, useless, or dangerous are faced with an interesting problem, one similar to the person Freud mentions somewhere in his analyses of humor. This person claimed that he not only did not believe in ghosts, but that he wasn't afraid of them either. Rejecting the dichotomy between the learned and the innate and accepting interactionism as the only solution seems to be a good example of the conceptual confusion. Hailman for example (and many others) rejects the instinctive learned dichotomy on the one hand and accepts the viewpoint of "continual interaction" on the other. One need only to ask *what* is interacting (ghosts) to plunge us back again into the dichotomy issue.

Now, we are interactionists too, and we start our analyses by investigating at the different levels of integration the *source of adaptedness*. There are numerous examples where ethologists have, by careful ontogenetic studies, disentangled the contribution of learning in the development of functional patterns, e.g., the copulatory grip of the polecat and the squirrel nut-opening, among others.

Ethologists certainly share the eager interest in developmental approaches. Since so much emphasis is put on the role of self-stimulation—a fact allegedly not seen by biologists—I want to mention that Roux's *Archiv für Entwicklungs-mechanik* was published first in *1894*. *Entwicklungsmechanik*, or in modern terms, *Entwicklungsphysiologie* has been a flourishing field ever since. It was Speman, approximately 50 years ago, who discovered that certain substances secreted at certain stages of the development from differentiated tissue trigger responses in the neighboring tissues. He called them *Induktoren*. Thus, the rim of the eyecup of the newt secretes a substance which induces the formation of a lens in the epidermis. If one transplants the eyecup into the belly region a lens will be induced in the epidermis of this region. But this is not to say that the information concerning the instruction to build a lens is contained in the organisor substance. The substance triggers a response which, as prospective potential, is given to any epidermis cell as an inherited capacity.

The study of behavior has revealed that phylogenetically evolved adaptations determine the behavior of animals in well-defined ways. Animals are provided with functional motor patterns at birth or at hatching. As soon as the newly hatched chaffinch emerges from its egg it shows the very characteristic food-begging response of gaping. A newly hatched duckling knows how to run, swim, dabble, and oil its feathers. Other behavior patterns mature during ontogeny without the need of acquiring patterned information concerning the specific adaptedness. The study of birdsongs provides examples (Konishi, 1964, 1965a, 1965b).

Besides such inborn motor patterns or skills, animals exhibit specific perceptual biases. They respond to certain environmental stimuli or stimuli configurations at first encounter, with specific motor patterns. Male sticklebacks raised in social isolation respond to a wax model of a conspecific with a red belly with fighting behavior. If the model shows a silvery, swollen belly it gets courted. During the breeding season these features characterize a rival or a mate respectively. The capacity to respond to certain signals without prior conditioning is based upon detector devices tuned to certain stimuli. Since by means of these mechanisms behavior patterns are released upon the reception of adequate key stimuli, they have been called releasing mechanisms. The signals evolved in mutual adaptation to fit them in turn are called social releasers (Tinbergen, 1951; Cullen, 1968).

Animals, furthermore, are known to be motiviated by the inbuilt physiological machinery known as *drives*. The term is to be understood as descriptive for the fact that animals are not passively waiting for stimuli to impinge upon them to give response. There is no unity drive mechanism; however, a diversity of physiological mechanisms act to spur an animal to seek in what is called appetitive behavior for stimulus situations that allow specific motor patterns to be performed (Lehrman, 1955; Hinde, 1966; von Holst, 1935). Sexual behavior, hunting, feeding, drinking, and, at least in some species, aggressive behavior are in part based upon such internal motivating mechanisms.

Finally, learning is determined by phylogenetic adaptations in such a way that animals learn what contributes to their survival and change their behavior by experience adaptively. It has been found that some animals learn during sensitive periods to attach certain reactions to objects, and once they are conditioned to them, the fixation seems resistant to extinction, to the extent of irreversibility in some cases. This phenomenon was called imprinting (Lorenz, 1935; Hess, 1973; Immelmann, 1966).

Learning disposition, similarly to drive, is a descriptive term and does not at all imply a unitary mechanism. The study of birdsong indeed clearly demonstrated that the same result—in this case that a bird learns the song of a conspecific—can be achieved in different ways (Konishi, 1964, 1965a, 1965b, Marler, 1959, Thorpe, 1961).

Chaffinches, for example, know what they have to imitate. From a variety of tapes offered to them, they preferably pick the species song for imitation. They know by means of an innate schema—Konishi coined the term *template*—which is the right song for imitation. In the zebra finches the learning of the right song is normally secured by a period during which the animal is particularly sensitive to songs it hears. What is memorized during this time has priority over succeeding experiences. Motivations, too, can provide a basis for an inborn learning disposition. We know that deaf-born children start to babble spontaneously.

It is easy to see why these discoveries of ethologists aroused the interest of those concerned with the study of human behavior. This field was until recently dominated by environmentalism. Extreme proponents of this philosophy have up until now argued that man is born as a blank sheet and programmed by learning during his ontogeny. Recent investigations, however, have proved that phylogenetic adaptations preprogram human behavior in ways similar to those of animals.

Thus studies of the deaf and blind born revealed a basically normal repertoire of facial expressions. Although the children could never see or hear a human being performing, they laugh and cry, smile, and show the complete anger syndrome, to provide a few examples. It could be argued that they informed themselves with their sense of touch. I had the opportunity, however, to study deaf and blind born thalidomide children who had not even the opportunity to explore their environment with the help of their sense of touch.

The argument remains that shaping could have occurred, for example, by the mother rewarding every smile with friendly fondling and every crying with comforting. Such reinforcement can be expected to occur but it must start with already recognizable patterns of facial expression for the mother to respond. For the more complicated pattern of expression, like anger behavior, it is difficult to conceive how this could be determined by accidental shaping.

Deaf and blind born also show some basic types of social responses, the fear of strangers being of particular interest. Although these children never experience any harm from strangers, they discriminate by their sense of smell between familiar and unfamiliar persons. The latter release the fear response. The child withdraws and seeks contact with a reference person. At a later age the fear-of-stranger response turns into active stranger repulsion. The child may act aggressive, pushing the stranger away before withdrawing. This response can also be observed in children of many different cultures. Man's inclination to live in exclusive groups and to show suspicion or even hostility toward strangers seems to be based on this inborn disposition.

The information we can get from the study of the blind born is limited. It is, however, of great theoretical importance, since the most important channels by which social behavior is normally released are blocked. Experiments with blind born help to get additional information. By verbal manipulation more complex responses—for example those of coyness and embarrassment—can be released. Another promising strategy of ethological research is the comparative method, including the cross-cultural comparison, a method we will discuss shortly.

Cultural Adaptations

In man, cultural evolution has, to an important extent, taken over biological evolution. Functionally, both are similar insofar as information is acquired and stored, but the mechanism of information storage and transmission differs. During cultural evolution information is stored in the brain of the individual by memory or by means of culturally invented artificial storage organs (books and electronic devices).

The basic invention that sped the cultural evolution of man undoubtedly was the development of language, which allows man to pass on instructions verbally, without the need to demonstrate the manipulation discussed or show the object under discussion.

In some primates it was observed that conspecifics instruct each other in certain skills. In one group of Japanese macaques a young female learned accidentally to wash the sweet potatoes which they were fed. Others learned from her and the skill spread, but each, of course, had to see another member of the species perform. Man can simply instruct another man: "Potatoes should be washed before eating!" With the invention of writing he need not even tell this personally, but can entrust the instruction to this code and preserve it for anyone in the future to be informed.

With cultural evolution, the slow procedure of genetic evolution is replaced by a much faster-working process. Cultural evolution is a process of adaptation which allowed man to adapt rapidly to a changing environment and to undergo culturally an adaptive radiation comparable to the adaptive radiation by specia-tion. Indeed the parallels are so striking that Erikson (1966) spoke of cultural pseudospeciation.

The parallels between cultural and biological evolution, however, go much further. Functionally, culturally, and phylogenetically evolved patterns of be-havior form a supporting framework for our behavior, which takes the burden of repetitive decision-making and probing from us. The patterns passed on have proved their selective value and to retain most of them is advantageous. And indeed, we value the beloved customs. We feel comfortable with them. On the other hand, we are adventurous, to a certain extent seeking new ways and methods. Conservatism preserves the patterns that have proved their value, while on a smaller scale the new is tried out and incorporated if of proven advantage.

Since it is extremely improbable that from one generation to the next the cultural inheritance of a group should be in totality of no further use, a culture would endanger its existence were it progressive to the extent of throwing its cultural heritage overboard. The fear of leaving the old customs is thus to be understood as a mechanism of positive selective value, as is the progressive search for the new. Both secure cultural evolution.

If one compares behavior patterns evolved during phylogeny with those cul-turally developed, one is struck by the numerous similarities in principle. They are particularly striking if we examine phylogenetic and cultural rituals. Both serve the function of communication, they serve as signals and are therefore similar in principle. A signal must be conspicous, unmistakable and simple at the same time. The signaling color patterns on the body sides of coral fishes, which serve for species recognition, and the flags by which ships signal their na-tionality indeed share these features. If the signals are motor patterns (expressive movements) they undergo parallel changes during their evolution to signals. Regularity, repetition, mimic exaggeration of the amplitude of movement and the like can be observed in cultural as well as in phylogenetic ritualization. Both processes start with patterns existing as preadaptations. Cultural rituals of bond-ing submission or threat have much in common with phylogenetic rituals serving the same function. Weapons are demonstrated in intention movements of attack as a threat display or turned away in submission or during certain types of greet-ing, as mentioned above. Presents are given to the host, in an intent to bond. There is little cultural variation in the basic theme.

The laws determined by functions that shape phylogenetic adaptations are operating in cultural evolution as well. The forces of selection remained the same. We may eventually hope to be able to extrapolate from the study of phy-logenetic ritualization the probable course of cultural evolution.

This is not to say that the similarities in human and animal rituals are solely to be explained by the same laws of function. Phylogenetic adaptations in form of perceptual biases may be responsible for the development of cultural rituals in man along particular lines. Since we respond to certain signals of the small child with friendly behavior they are often used in cultural rituals as infantile appeals to appease aggression and to bond at the same time. But the point is that even here where phylogenetic adaptations are not involved, similarities between cultural and phylogenetic adaptations are to be expected, since the selection pressures operating are the same in both cases.

CROSS-CULTURAL STUDIES IN HUMAN ETHOLOGY

According to what I have said so far, the aim of a cross-cultural study of human behavior is to understand the biological and cultural determinants of behavior and to explore the laws of function which govern the development of cultural as well as phylogenetic adaptations.

Cultures behave like species and can be compared as if they were. By applying the various criteria, rituals and languages can be traced as homologous. Analogies on the other hand, for example in the cultural patterns of social structure, reflect adaptations to specific environmental demands. The cross-cultural study may indeed reveal patterns of social behavior typical, let's say, for hunters and gatherers in semiarid zones, for pasturalists, farmers in temperate zones, and the like.

A special problem for the ethologists is posed by those patterns that occur as universals in every culture. Patterns that are inborn to man certainly should occur as universals, since man biologically still forms one species and inborn patterns of behavior are, to our knowledge, very conservative. This statement, however, cannot be twisted to say that universals are inborn. If certain factors of selection were to be the same everywhere, even culturally developed patterns should be expected to be universally the same, in principle at least. Finally, experimental factors in early childhood that occur universally may shape a behavior along similar lines in all cultures.

To explore these different possibilities thoroughly, cross-cultural studies with a multidisciplinary approach are needed. These include comparative studies on the ontogeny and studies on ecology and behavior. Most important is a documentation program by means of which behavior patterns are made available for comparison in an objective way.

I am engaged in such a program, filming unstaged social interactions and taping the verbal expressions cross-culturally. The filming is done with mirror lenses, so that people are not aware of being the focus of attention. In addition to every filmstrip, a protocol reports about the social context and about what the person did before and after the behavior filmed was exhibited, and finally what followed as a response. This is a prerequisite for later correlational analyses. I have dealt with this method repeatedly and extensively and do not intend to discuss the methology here in more detail.

Over the years I have collected a considerable amount of film documents of different cultures. At the same time, I concentrated on those cultures that are expected to change rapidly in order to rescue the data in time. Several and repeated visits were made to the Bushmen of the Kalahari, Waika Indians, several Papuan tribes, the Himba, two Central Australian tribes, and the Balinese, to mention but a few. In addition to these thoroughly documented tribes, I executed a number of pilot studies in other areas.

The documentation reveals that phylogenetic adaptations play an important role. The repertoire of patterns of nonverbal communication is alike in all the cultures visited so far, down to the minutest details, which is rather striking in face of otherwise contrasting cultural differences. The culturally traditioned codes of communication, in contrast, are apt to change rapidly in cultural evolution, as the evolution of languages demonstrates.

Coyness is expressed with the same patterns of behavior everywhere—patterns which are ritualized expressions of intention movements of approach (contact seeking) and withdrawal (contact avoidance). This ambivalance is expressed with the eyes by alternatively looking away with lowered lids (cutoff) and looking toward the partner who released the ambivalent behavior. In addition, body posture and movements express approach and withdrawal either in alternation (turning to and away, lowering the head and/or turning it away, and turning to face the partner) or in simultaneous ambivalence (looking at the partner but turning the body half sideways so that the shoulder is shown).

Hiding of the face behind the hand might occur as another means of cutoff, and it is interesting that I observed this covering of the eyes even in blind born. Often the tongue is pushed out of the mouth for a moment and laughing may occur.

Another universal pattern that I repeatedly described occurs during greeting encounters. When greeting from a distance, people were found to nod, smile, and if very friendly, to raise the eyebrow in a rapid movement for approximately $\frac{1}{6}$ of a second. The pattern is a ritualized expression of happy surprise (Ah, it's you!).

I filmed this eyebrow greeting in jungle-dwelling Indians on the Upper Orinoco, in Papuans, Samoans, Balinese, Europeans, Bushmen, Himba, Chinese, and others. Some differences do exist between different cultures but they concern merely the ease with which this behavior can be elicited.

Japanese consider it unseemly, but Samoans greet everyone in this way and even use it to underline affirmation. Europeans, roughly, stand in the middle of these two extremes. They do not greet all arrivals this way but use the signal generously with good friends and also when flirting and giving enthusiastic consent. The importance of this minute signal in everyday life may be seen from the attention women pay to their eyebrows, shaving and pointing them to make the movement more significant, often coloring the upper eyelid and the area just beneath the brows to serve as contrast marking. Yet, until recently, eyebrow flash was unknown to science. Poets spoke of it as a lighting up of the eyes, but in vain do we search for its description in texts dealing with human facial expressions. We could add a great number of examples to this list.

The cross-cultural comparison also reveals basic similarities in more complicated rituals. Thus feasts, despite their vast cultural differences, at first glance show structural characteristics which are universal. Guests and hosts at such occasions try to impress each other by a variety of displays. These themselves can vary culturally. Display of personal skills (for example, in a dance performance), ornamentation, showing off of armaments, or demonstrating wealth, even to the extent of destroying valued property (see for example the potlatch performance of the Kwaikutl), are all different means of display.

In individual behavioral expressions, inborn motor patterns can be observed and some of the displays are shaped by the inherent response mechanisms of the observers, for example, the emphasis of the shoulders in male ornamentation. To an important degree, however, the boasting is done by culturally determined means, although along functionally similar lines. The basic motivation to impress is always present and could well be inborn, and the common occurrence of competitive games could be derived from the same motivation.

But feasts are not solely a platform for display. Bonding is a basic function and it has its own motivation. There seems to be less variation in the expression of affection. One universal way is to demonstrate concern—remembering those that died recently by jointly wailing, or in a more culturally ritualized way, by expressing the concern verbally or by symbolically putting down a bouquet in front of a memorial grave. It is a universal habit to feed the guests, and of course, that they must be fed if the ceremony lasts any length of time is a functional explanation that could explain this. But this does not seem to be the whole story. Food plays such an important role that even those not attending often get their symbolic share and one also feeds spirits and gods to bring them into a friendly mood.

Food certainly has more than practical importance. It is used to make and keep friends, and indeed, dining and drinking together forms a strong bond. This is also true in many animals, where ritualized feeding plays an important role, as in courtship behavior. Here it is mostly parental feeding that became ritualized to courtship feeding or other bonding feeding. Accordingly, the partner gives infantile appeals before feeding and infantile reactions afterwards, although again in an often highly ritualized form.

In man we have this ritualized parental feeding as well; it is the kiss which clearly derived from kiss-feeding. We observe that very small children, before they are able to speak at all, establish bonds with strangers by giving them a morsel and taking one in return. Such dialogues of giving and taking I observed cross-culturally and they play a significant role in early childhood. For this reason I believe that phylogenetically evolved behavioral programs play an important role in bonding via food or in its derived form via the exchange of presents (for details see Eibl-Eibesfeldt, 1972a,b).

Common occurences during feasts are group activities, for example dances in which a group of people move together. Such concerted actions bond and at the same time they separate the groups by setting them off against others which, not being familiar with the rules of the ritual, are not able to keep pace in the dance. Only those that know the rules are "in." It is clearly a factor contributing to cultural pseudospeciation. The fact is that concerted-action bonding has its innate bases. We know that rhythms synchronize the heartbeat and other physiological processes, even in lower vertebrates. Clearly, march music functions that way, but a dance is a ritual of synchronization, too. The deep enjoyment small children derive (even before the age of one) from a concerted action and the fact that they grasp the principle immediately by imitating spontaneously, starting with peek-a-boo and going on to "ten little Indians," is a clear indication of the innate learning disposition involved.

Many of these synchronization rituals remind us of the antiphonic rituals in birds whereby each bird sings some stanzas alternately. Since both are perfectly synchronized, they sing one melody and it is difficult for the observer to realize that actually two birds are singing together. Rituals of that type can be observed at a very early age between mother and child.

The same structural elements occur in feasts all over the world and the basic structure remains the same, although the emphasis can shift to concentrate on some features in particular, depending on whether the feast serves to strengthen a village alliance, whether a new family bond is established at a marriage, or whether the occasion is a mourning ritual.

Of particular interest is, furthermore, the fact that at certain encounter situations basically the same events can be observed in very small children. If a person visits a family which has a small child of about one to one-and-a-half years of age, he may observe that after a while the child loses his initial fear and starts to relax. He then takes initiative to contact the stranger, often by showing him one of his favorite toys and even handing it over, expecting to get it back, however. Some offer a morsel of food and brighten up when the stranger eats it and offers a morsel in return. The child will furthermore put itself into the focus of attention by various displays of skills, running about, climbing things; in short by showing off. And it will finally invite the other to join in its actions.

Phylogenetic and cultural adaptations intercalate in human behavior in a most intricate way. To disentangle the inborn from the culturally traditional is possible by cross-cultural studies of behavior, including the study of ontogeny. The task is complicated by the fact that the course of phylogenetic and cultural evolution is determined along similar lines by the forces of selection.

CULTURAL AND BIOLOGICAL RITUALIZATION OF AGGRESSIVE BEHAVIOR

The analogies between cultural and biological evolution can be impressively demonstrated when one compares the different biological and cultural ritualizations of aggressive behavior.

In contrast to interspecific aggression which aims at the destruction of the opponent, be it predator or prey, intraspecific aggression is commonly controlled in a way that no serious physical damage of the opponent results. There are some examples which would seem to violate the statement, but the species either have no serious armaments or rarely find themselves in a situation which leads to the destruction of the opponent. For example, the capacity for flight may be so highly evolved that the loser can successfully retreat after the exchange of a few bites, whereas the winner is motivated not to follow but to stay in his territory. This means that the consummatory situation is not the opponent being damaged but being "not there." This holds true for a number of rodents.

Wherever there is real danger, however, that a conspecific might be killed during an encounter, special mechanisms have been evolved to prevent this. Poisonous snakes are known to wrestle according to rigid rules and never to bite each other. Rattlesnakes, for example, raise the front third of their body and try to hit the rival with sideways blows of their heads.

Marine iguanas, which could easily wound each other with their sharp tricuspid teeth, do not use them during fighting. If during the breeding season a rival male enters the territory of another, the territory owner will first display. He will orient his flank to the opponent and walk stiff-leggedly up and down, his dorsal crest fully erected. He will furthermore open his mouth wide, as if in intention to bite, and nod his head. If the rival does not retreat, a fight will follow. The opponents will turn toward each other and attack, but never biting, simply batting with their lowered heads. Each will try to push the opponent from the arena. This wrestling may last for some time until one realizes that he has little chance to win; he then assumes a submissive posture by lying flat on his belly. The winner will stop fighting immediately in response and wait until the loser slowly retreats. Examples of such tournaments are known in numerous vertebrates from fish to mammals. They show that two selection pressures have exerted their influence. One worked in favor of sparing the conspecific, and selective advantage is easy to visualize. A group in which individuals, due to lack of inhibitions, kill its group members would be at a disadvantage in competing with groups in which no intragroup-member killing occurred. On the other hand the fact that complicated rituals of fighting have evolved can be taken as proof for a strong selection pressure toward intraspecific aggression as a mechanism of spacing, otherwise it would have been the simpler solution to drop aggression completely where it was a source of danger for the species member.

Often, intraspecific fighting commences with the exchange of bites, but special submissive behavior stops the fight. In dogs, for example, the loser rolls on his back, in a posture that has been derived from infantile behavior (puppies expose themselves this way to the mother to be cleaned). Indeed, the loser in this position often urinates a little. This causes the winner to clean him, and then the loser, by wagging his tail, turns on patterns signaling friendly intent. And indeed what started as a fight may continue with friendly play. As I emphasized in a previous monograph on the mechanisms of bonding, this capacity of the dog not just to appease by turning away aggression-eliciting stimuli, but to bond in addition is an achievement which developed within the vertebrates as a consequence of parental brood-care behavior. Reptiles do not have this capacity to bond.

If we examine the behavior of man, we will encounter phylogenetic adaptations which in a similar way serve the function of aggression control and bonding just as in the previously described examples. A worldwide expression of submission and cutoff is pouting, in combination with lowering the head, gaze aversion, and a slight turning away. Bowing and haddling may be related to headlowering and are certainly stronger acts of submission which in cultural elaboration may go as far as prostration. Another pity-releasing, and thus aggression-inhibiting, signal is crying. Bonding and aggression-inhibiting patterns are furthermore readily observed in greeting encounters the world over, for example, the nodding eyebrow-flash with head-toss and finally smiling.

Above and beyond his ability to appease, man is strongly motivated by a drive to bond. That this disposition is rooted in the mother–child relationship can be derived from the fact that mother–child signals serve in bonding between adults (Eibl-Eibesfeldt, 1972a, b). Biologically, man undoubtedly is preprogrammed for a friendly, cooperative life in groups. In fact, certain observations indicate that his urge to bond even with a stranger is often stronger than his urge to fight. How could one otherwise explain the fact that soldiers engaged in trench warfare so often stop fighting and exchange cigarettes instead, a phenomenon cynically called demoralization of the troops. It is for this reason that warring parties must be constantly indoctrinated by propaganda which instructs one not to fraternize, since the members of the other side are not real human beings. Once a communicational barrier is established, bonding and appeasing signals are no longer effective.

However, with the invention of weapons the situation became critical, even in cases of aggression between members of the same group. A rapid blow with an axe carried out in a fit of rage can kill the opponent without giving him any chance to send appeasing signals. Thus, for such reasons our inborn patterns of aggression control became ineffective, except perhaps within the family. Cultural patterns had be be developed to secure peaceful coexistence in a group. We have already mentioned greeting rituals in which the normally carried weapons are presented to the host or taken off. Fighting itself became ritualized in many ways, for example in the Iko Bushmen of the Kalahari. If a person is insulted he will complain in the evening when sitting before his fire. Everyone will hear him and, although courtesy demands that names are not pronounced, everyone knows of course to whom his complaints are addressed. One can be sure that the addressee will come the next morning to speak "soft words" in an effort to make the peace. Thus fighting is avoided.

The Eskimos are also renowned for their ritualized ways of fighting. The tribes of Siberia, Alaska, Baffinland, and Northwest Greenland solve their disputes by wrestling. The Central Eskimos (Hudson Bay, Bering Strait) ear-cuff each other. The tribes of West and East Greenland, of the west coast of Alaska and from the Aleutian Islands solve their quarrels by alternatively singing stanzas in which they ridicule the rival. The audience then decides who wins. This is not to say that Eskimos are not "aggressive." Members of one group have even been reported to kill strangers who have violated their rules.

Where weapons are used, the way of using them has often become ritualized. Waika Indians perform chest-pounding duels which may escalate into club duels in which the opponents alternatively hit each other on their head. It is the rule that only the shaved part of the head is hit. The clubs are long, made from hardwood, and they have sharp edges which cause gaping wounds. After one had delivered his blow he has in turn to offer his head to receive a blow. This continues until one gives up or is simply not in a position to deliver his blow. The Waika warriors are proud of the scars which furrow their shaved heads. A clubbing duel between two villages may prevent a shooting war and even initiate reconciliation, since both have worked their anger off and proved to be men (Chagnon 1968, Biocca 1970).

Comparable duels are known from Europe. They differ locally but those who are familiar with the rules of conduct are protected from fatal escalations of aggression. However, as we have said already, man has the tendency to pseudo-speciate culturally. Culture-specific patterns evolve, including language, and serve as barriers demarcating the ethnic groups. Communication is hampered or even effectively prevented, particularly once language barriers evolved. Conflict situations then no longer have the character of individualized aggression. They have become remote from the biological level of aggression moving to the cultural level of war. And war, by pattern and function, equals interspecific aggression, aimed at destruction of the opponent, and this is increasingly true the more anonymous the groups confronting each other become. Cultural pseudospeciation leads to war, which is a cultural phenomenon.[1] Biological aggression controls, as well as group-specific modes of cultural control, do not suffice anymore to cope with war, and new controls have to be developed culturally, always provided, of course, that we want peace to come, that we still consider mankind as belonging to one species *despite* the cultural diversity. A moral decision has to be made.

Were we to decide to promote cultural subspeciation by erecting barriers against communication, wars would continue until one group achieved domination. However, even then, world peace would not necessarily be established, since the mechanism of subspeciation would continue to operate. But we have another choice to which we are already preadapted. As elaborated in the previous chapters, all men share phylogenetically inherited behavior patterns and motivations. This equipment provides a basis of reference. Only due to these inborn behavioral programs are we capable of understanding each other on a basic emotional level, which means understanding each other on a basic emotional level, which means understanding in the real sense. If we still experience the conflict in killing a conspecific when we make war against people of other races and cultures, it is due to this shared inheritance. Against the tendency of cultural seclusion, the idea of a brotherhood of men is a strong uniting force bonding men of different religions and cultures. This does not necessarily imply complete cultural amalgamation. Religions and political ideologies have proved to bond people and still allow cultural diversity to flourish.

That our hopes for world peace is not merely a philanthropic utopia can be observed from a number of facts. Throughout history it can be observed that even warfare tends to be controlled by rituals that prevent utter destruction. It is possible to surrender with a good chance to survive. The internationally agreed upon code of conduct demands that the captured soldier is not killed, that civilians are not maltreated by the occupation troops. Many acts are considered as war crimes which go against this code. It is interesting to note, however, that mutual agreement is necessary for the functioning of such codes. That a difference could be observed in the treatment of Russian, German, and British prisoners in World War II is a clear example of this.

So it seems that analogues to the ritualization of aggression in warfare takes place on the level of cultural evolution which hopefully may lead to tournament-like struggles on the stage of a world parliament. Such a development demands a

---

[1] One may argue that civil wars are particularly cruel, for although the parties belong to the same people and therefore understand each other's code of conduct, all efforts are put into demarcating one from the other. This is clear from the campaign of dehumanization which both sides engage in.

clear ethical decision. We are emotionally preprogrammed for humanity by in-born mechanisms of bonding and aggression control. If we follow our inclina-tion, peace will become inevitable. Of course we have to realize that the func-tions of war must have substitutes: for example, making resources accessable or controlling population growth. These problems have to be solved by measures based on reason. The first signs of such a development can be seen, although man so far has not done very much to steer his fate but exposes himself quite passively to the forces of selection.

At the end of his book *On Aggression*, Konrad Lorenz promotes the thesis that man is not good enough for life in the modern large societies. By reason, he can accept the demand to love his fellow man, but this capacity does not suffice. Man can only feel love for those personally known to him. Our goodwill cannot change this situation, only evolution can bring change about. He trusts that se-lection will breed into man the capacity to love all men indiscriminately. Modern man is considered to be the long-searched-for "missing link" from ape to real man. Also, E. von Holst (1969) expressed the opinion that the peaceful man has yet to be bred.

I hope that I have been able to show that my confidence in man's future is not merely based on wishful thinking but on possibilities deduced from the observed ways of phylogenetic and cultural evolution. Phylogenetic adapta-tions in our motivational structure predispose us to overcome the communi-cational barriers and will help to achieve world peace as a result of cultural evolution.

We are, of course, aware that the inclination to subspeciate and to form ex-clusive groups which counteracts friendly bonding beyond such groups is strong. But we can detect from history that man in principle is capable of forming anonymous groups which are still bonded effectively, i.e., on an emotional basis. By symbol-identification and by extending the family ethos to encompass all members of the town, tribe, and/or national community, human groups grew and found unity. There is no reason to assume that these mechanisms could not operate to unite men on the level of mankind. Certainly the tendency to see brothers in all men is fostered by the modern media. It is difficult to tell people today that those living in other parts of the world are not people, when televi-sion informs them daily that all men are troubled by the same problems, striving for happiness along similar lines and, in short, behave basically alike. Certainly we have to be aware that this uniting tendency is counteracted by the tendency of subspeciation, but it is important to see that we can intellectually decide which of our dispositions to follow. We can decide whether to look at mankind as one family and live and plan accordingly or compete ruthlessly, split up into numerous ethnic groups acting as if they represented different species. The future cultural evolution will decide the fate of man. Building upon our in-herited dispositions will eventually lead to their full realization—whether for the good or for the bad is up to us to decide.

DISCUSSION

From a number of recent discussions it can be seen that some basic mis-conceptions about the concepts of ethologists do exist. It is argued, for example, that anything inborn to man has accordingly to be accepted as his fate, and that ethologists, by pointing to the inborn, would provide material for those who teach the basic unchangeability of human society. Ethologists are also accused of being exculpative, i.e., when pointing out phylogenetic adaptations that deter-mine aggressive behavior. Thus Lumsden (1970) writes:

> The danger with the "instinct of aggression" theory is, that far from emanci-pating man, it may enslave him to a reactionary ideology by apparently demonstrating the "biological necessity" of an authoritarian social system organized for internal and external repression. (p. 408)

In Montague's collection of polemics (1968) one may find similar "argu-ments." It has repeatedly been pointed out by ethologists that man is a cultural being by nature, and we also emphasized that cultural evolution is continuing the biological evolution in man. By means of cultural patterns man is not only in a position to check inborn behavior and motivations but in order for functional behavior to occur he must necessarily do so. Whereas in animals long chains of behavioral events are preprogrammed down to small details, in man this is not so. We are motivated by inborn drives and certain inborn norms control the events, but only in a fairly wide frame.

Cultural control confines behavior within more limited boundaries. Since these cultural patterns can vary from place to place, man was able to open for him-self a multitude of ecological niches. After all, an Eskimo needs different controls for his aggressive or sexual impulses than, say, a Masai or a modern urban dweller.

In addition we can modify our cultural control patterns for behavior when it seems necessary. Of course by emphasizing man as the cultural being, ethologists were promptly accused of advocating repressive education. Propagators of the nonauthoritarian education claim that a human being should be left to develop from within himself. But on what bases? Following our own natural tendencies? Though we are aware that our behavior is still, to a large degree, controlled by phylogenetically derived patterns we would nonetheless not assume that these alone would suffice.

It is odd, indeed, that those who place such emphasis on the role of the environment with respect to the shaping of the human personality do not want to pass on the cultural guidelines evolved for the control of our behavior and by fairly thoughtless experimentation endanger the development of the individual. Students of child behavior are very well aware that much of the child's aggres-sion is not all caused by frustration but used instrumentally as a means of social exploration. The child wants to learn about the limits of toleration; it wants to get a response. If it does not get an answer, its behavior is apt to escalate.

It should be clear therefore, that ethologists accept nothing as an inevitable fate. Where the study of behavior has proven that inherited patterns have lost their originial adaptedness in modern society and are carried along as burdens of history, education will provide the means of control. Insight into the physiological mechanisms as well as the understanding of the original function will help to develop the adequate strategies. Since Lorenz, in connection with his studies on aggression, was often accused of fostering apologetic and fatalistic attitudes, I may quote a passage from his repeatedly attacked book.

> With humanity in its present cultural and technological situation, we have good reason to consider intra-specific aggression the greatest of all dangers. We shall not improve our chances of counteracting it if we accept it as something metaphysical and inevitable, but on the other hand we shall perhaps succeed in finding remedies if we investigate the chain of its natural causation. Wherever man has achieved the power of voluntarily guiding a natural phenomenon in a certain direction, he has owed it to his understanding of the chain of causes which formed it. Physiology, the science concerned with the normal life processes and how they fulfill their species-preserving function, forms the essential foundation for pathology, the science investigating their disturbances. (1966, pp. 29–30)

Lorenz sees the importance of compensating for what is nowadays maladaptive behavior. The final adaptations, he argues, will come from evolution. Since war, however, is a cultural phenomenon, cultural evolution has to bring about its adequate control. By our inherited background we are already *biologically* fit for life in large societies; it is a lack of *cultural* adaptations that causes problems. The ritualization of warfare has just begun and the efforts to unite as a family of men are still counteracted by tendencies of subspeciation.

Let me emphasize once more, ethologists agree that man's problems can be solved by education. In contrast to the extreme environmentalists, as represented, for example, by Skinner, we derive our strategies by studying the nature of man. It is my conviction that this consideration is in the end more humane than strategies which are based on ideology alone. Thus, Skinner proposed to shape man for his survival by the technique of conditioning. The guiding norms have to be derived functionally. What contributes to the survival of a culture must be accepted. According to Skinner, no inborn norms exist. It is easy to see where such an extreme cultural relativism leads to. If a culture agrees upon euthanasia on practical grounds, no one could argue against it. Fortunately, it can be shown that binding ethnical norms, universal for men, are part of our inborn nature. Whoever denies the existence of human nature runs into the danger of constantly frustrating man by overburdening him without need.

Repeatedly, we pointed to the fact that cultural evolution is shaped by selection along lines similar to the phylogenetic evolution. I propose therefore the hypothesis that the study of phylogenetic evolution of behavior provides models for cultural evolution and, given a set of conditions, we should be in the position to predict the possible pathways of cultural evolution. This may enable us to overcome the blind trial-and-error mechanisms by analyzing the selection mechanisms which have governed our cultural evolution so far and thus to start a cultural evolution guided by reason.

SUMMARY

Phylogenetic adaptations determine human behavior in various ways. Man has a repertoire of inborn motor patterns and innate releasing mechanisms enable him to respond without need of conditioning to certain stimuli situations with a set of given responses. Releasers evolved as signaling devices. Motivating mechanisms activate man in a similar way to animals and inborn learning dispositions determine what and when something is preferably learned, and when. Thus man is to a certain extent preprogrammed to act.

Cultural evolution, however, has taken over in man and has been substituted for the biological one. Functionally, both are comparable as they are fundamentally the same. Both can be seen as processes of information acquisition and information storage, by which adaptation results. The mechanisms by which this takes place, of course, are different in both cases. In both processes, conservative and progressive elements interact in a balanced way, ensuring survival. Since the same selection pressures operate during biological and cultural evolution, the resulting patterns of adaptations are shaped along similar lines. Knowledge of the laws forming phylogenetic evolution may therefore help in our understanding of cultural evolution, in extrapolating its probable future course, and eventually in freeing ourselves from the game of chance by which evolution has worked so far and, by insight, guiding our future cultural evolution. To understand the laws determined by function is one aim of animal and cross-cultural comparison. To determine the extent of phylogenetic preprogramming is the other important task for human ethology. Environmentalistic philosophies, in lack of basic biological knowledge and partly blinded ideologically, have failed to take these facts into consideration which led to a cultural relativism, with dangerous totalitarian consequences as far as the strategies of education are concerned.

REFERENCES

Biocca, E. Yanoama. *The narrative of a white girl kidnapped by Amazonian Indians*. New York, E. P. Dutton, 1970.

Chagnon, N. Yanomamö. *The fierce people*. New York: Holt, Rinehart and Winston, 1968.

Charlesworth, W. R. Limits to the application of ethology to problems in human development: A real problem? In press.

Cullen, E. Experiments on the effects of social isolation on reproductive behaviour in the three-spined stickleback. *Animal Behavior*, 1968, *8*, 235.

Eibl-Eibesfeldt, I. *Nannopterum harrisi* (Phalacrocoracidae): Brutablösung. Encycl. Cinem., E 596. Publ. zu wiss. Filmen, 2A, Göttingen: Inst. wiss. Film, 1965.

Eibl-Eibesfeldt, I. *Ethology, biology of behavior*. New York: Holt, Rinehart and Winston, 1970.

Eibl-Eibesfeldt, I. *Die Iko-Buschmanngesellschaft*. Aggressionskontrolle and Gruppenbindung. Monographien zur Humanethologie I. Munchen: Piper, 1972*a*.

Eibl-Eibesfeldt, I. *Love and hate*. New York: Holt, Rinehart and Winston, 1972*b*.

Erikson, E. H. Ontogeny of ritualization in man. *Philosophical Transactions, Royal Society of London B*, 1966, *251*, 337–349.

Hailman, J. P. The ontogeny of an instinct, *Behaviour (Leiden), Suppl. 15*, 1967.

Hess, E. H. *Imprinting: early experience and the developmental psychobiology of attachment*. New York: Van Nostrand, 1973.

Hinde, R. A. *Animal behavior, a synthesis of ethology and comparative psychology*. New York/London: McGraw-Hill, 2nd ed., 1972.

Holst, E. v. Über den prozes der zentralen koordination. *Pflüg. Arch.*, 1935, *236*, 149–158.

Holst, E. v. Zur Verhaltenphysiologie bei Tieren und Menschen. I. und II. Munchen: Piper, 1969.

Immelmann, K. Zur Irreversibilität der Prägung. *Die Naturwiss*. 1966, *53*, 209.

Konishi, M. Effects of deafening on song development in two species of Juncos. *Condor*, 1964, *66*, 85–102.

Konishi, M. Effects of deafening on song development of American robins and black-headed grosbeaks. *Z. Tierpsychol.*, 1965*a*, *22*, 584–599.

Konishi, M. The role of auditory feedback in the control of vocalisation in the white-crowned sparrow. *Z. Tierpsychol.*, 1965*b*, *22*, 770–783.

Lawick-Goodall, J. Van. The behavior of free-living chimpanzees in the Gombe Stream Reserve. *Anim. Behaviour Monogr.*, 1968, *1*, 161–311.

Lehrman, D. S. A critique of Konrad Lorenz's theory of instinctive behaviour. *Quart. Rev. Biol.*, 1953, *28*, 337–363.

Lehrman, D. S. The physiological basis of parental feeding behaviour in the ring dove (*Streptopelia risoria*). *Behaviour*, 1955, *7*, 241–286.

Lorenz, K. Der Kumpan in der Umwelt des Vogels. *Journal of Ornithology*, 1935, *83*, 137–413.

Lorenz, K. Phylogenetische Anpassung und adaptive Modifikation des Verhaltens. *Zeitschrift Tierpsychologie*, 1961, *18*, 139–187.

Lorenz, K. *On aggression*. New York: Harcourt, Brace and World, 1966.

Lumsden, M. The instinct of aggression: Science of ideology? Futurum, *Zeitschrift für Zukunftstforschung*, 1970, *3*, 408–419.

Marler, P. Developments in the study of animal communication. In P. R. Bell (Ed.), *Darwin's biological work*. Cambridge: University Press, 1959, pp. 150–206.

Montague, M. F. A. *Man and agression*. New York: Oxford Univ. Press, 1968.

Skinner, B. F. *Beyond freedom and dignity*. New York: Knopf, 1971.

Thorpe, W. H. Bird song. The biology of vocal communication and expression in birds. *Cambridge Monographs in Experimental Biology*, 1961, *12*.

Tinbergen, N. *The study of instinct*. London: Oxford University Press, 1951.

Tinbergen, N. Einige Gedanken über "Beschwichtigungsgebärden." *Zeitschrift Tierpsychologie*, 1959, *16*, 661–665.

Wickler, W. Vergleichende Verhaltensforschung und Phylogenetik. In: G. Heberer (Ed.), *Die Evolution der Organismen*. Bd.1, Stuttgart: Fischer, 1967, pp. 420–508.

Wickler, W. *Verhalten und Umwelt*. Hamburg: Hoffmann und Campe, 1972.

Unpredictability in the Etiology of Behavioral Deviations

JULES H. MASSERMAN

There is an aphorism that is applicable to scientists from Newton to Einstein: namely, that the vistas we survey were rendered visible only because we stood on the shoulders of giants—whom we soon forget. How infrequently we mention Fabre or Von Fritsch when we're talking about ethology. In neurophysiology, Walter Cannon is similarly neglected. My own greatest teacher, Adolph Meyer, is rarely mentioned by psychiatrists. Fortunately, in Russia this is not true. When I visited the Neuropsychiatric Institute in Moscow, there was hardly a speaker—Snezhnevsky, Lebedev, Schirina, Vartanian—who did not pay tribute to the aged Dr. Anokhin, who sat beside me. May I suggest that we here pay similar tribute to Academician Anokhin, who should have been here in my place?

As to my own presentation, I shall presume, with a bit of hubris, that some of you may have read some of the things I've published. I make this presumption because in the indices of the books written by members of this distinguished audience, I often see my name, Masserman—neatly sandwiched between masochism and mass hysteria.

With this as a possible background, I shall review only those of my findings and inferences that are relevant to this Colloquium, and even then only in summary form. But first a passing reference to our primary interests as clinicians.

As scientists, we physicians expect our medical disciplines to have a logical

Jules H. Masserman ● Northwestern University Medical School, Evanston, Illinois.

development, to be continuously refined by observation and experiment, and to stand the test of operational predictability. Then, possibly to spare the sensibilities of colleagues whose specialty does not always seem to fulfill these criteria, we concede that "much of psychiatry is not a science but an art." As a psychoanalyst with at least partial scientific respectability as a neurologist, I accept this stricture, taking comfort in the fact that the Oxford Dictionary defines art, too, as "skill derived from knowledge and experience." But I submit that we are also evolving concepts of the etiology and therapy of behavior disorders that are historically consistent, clarify and integrate previous theories, and conform to the third criterion of a science: experimental demonstrability. Since, for the last quarter century, I have been particularly concerned with the third heuristic source, permit me to review some of our relevant animal experimental work before returning to our ultimate clinical interests.

ANIMAL EXPERIMENTS IN THE DEVELOPMENT OF "NORMAL" AND "ABNORMAL" BEHAVIOR

Significance

Two preliminary questions still sometimes arise at this juncture—although, Pavlov, Darwin, and Heaven be thanked, far less often than formerly—to wit: "But are animal experiments *really* [sic] applicable to the subjective and social complexities of human behavior?" Or, in a more subtly patronizing vein, "Are you not *anthropomorphizing* your data and your inferences?"

In briefest riposte, neurophysiologically, just as the human central nervous system, although undoubtedly more highly developed in some respects, is nevertheless of the same basic design as that of other chordates, so also is human behavior more highly dependent on complex communicative and social influences, and thereby more contingent and versatile—*but not different in basic adaptational patterns*—than the behavior of man's somewhat less pretentious, preemptive, pompous, and pugnacious fellow-creatures.

Epistemologically, no "datum," whether labeled "material," "experimental," "intuitive," or whatever, is ever really "given" in pristine purity by a gracious cosmos; instead, since data can never be more than man's incomplete and exceedingly fallible perceptions of the universe, terms such as *real*, *objective*, *subjective*, *anthropomorphic*, etc., become tautologic shibboleths, meaningless to the modern logical-positivist rationale of science. Being human, some psychiatrists will continue to take a peculiar pride in the supposed uniqueness of human neuroses, but even they may find the behavior of animals disconcertingly reminiscent not only of the development of individual and social conduct in man, but significant to the therapy of his neurotic and psychiatric deviations.

Biodynamic Principles

But one more task remains before we proceed to experiment: the clear formulation of parsimonious hypothesis[1] in testable form—a practice for which psychoanalysts have not been noted. In a not altogether successful attempt to supply this deficiency, I have essayed to test in our own work the following four relatively simple biodynamic principles, respectively applicable to motivation, learning, adaptability, and neurotigenesis, and putatively operative in both animal and human behavior.

Principle 1. Motivation. The behavior of all organisms is actuated by physiologic needs, and therefore varies with their intensity, duration, and balance.

Principle 2. Perception and Response. Organisms conceive of, and interact with, their milieu not in terms of an absolute "external reality," but in accordance with their individual capacities, rates of maturation, and unique experiences.

Principle 3. Range of Normal Adaptation. In higher organisms, this makes possible many techniques of adaptation, and this versatility in turn renders the organism capable of meeting stress and frustration and maintaining an adequate level of satisfaction by (a) employing new methods of coping with difficulties when the old prove ineffective, or (b) by modifying objectives or substituting new ones when the old become unattainable.

Principle 4. Neurotigenesis. However, when physical inadequacies, stresses of uncertainty, or motivational–adaptational conflicts exceed the organism's innate or acquired capacities, internal tension (anxiety) mounts, neurophysiologic (psychosomatic) dysfunctions occur, and the organism develops overgeneralized patterns of avoidance (phobias), ritualized behavior (obsessions and compulsions), aberrant social transactions, and regressive, hyperactive, aggressive, or bizarrely "dereistic" (hallucinatory, delusional) responses corresponding to those in human neuroses and phychoses.

These inferences, always subject to modification by further data, have emerged from the work conducted by my associates and myself during the past three decades. Permit me very briefly to review our results.

---

[1] Said William of Occam, sharpening his famous Razor: "*Essentia non sunt multiplicanda praeter necessitatem.*" Added Albert Einstein, after thirty more generations of scientific thought: "The grand aim of all science is to cover the greatest number of empirical facts by logical deductions from the smallest number of hypotheses or axioms."

ONTOLOGIC STUDIES

Phases of Development

Our records and films of individual animals of various species from infancy to adulthood have confirmed the principle that the young of all organisms normally evolve through an orderly succession of stages during which sensory modalities are distinguished and resynthesized, integrated concepts of the environment are developed, manipulative skills are refined, early dependencies on parental care are relinquished in favor of exploration and mastery, and peer and sexual relationships are sought through which the animal becomes normally "socialized" in its group.

Formative Experiences

Young animals given opportunities for continuously nutritive and protective contacts with parents or their surrogates, and later with peers, manifest exploratory self-confidence, acquire motor and interpersonal skills and develop "social acculturations."

Learning

The growing infants show patterns of dependency, exploration, play, fetishism (i.e., attachment to objects such as blankets representing early securities) tantrumy rebelliousness, gradually more effective sexual techniques, and other characteristics significantly parallel to those in human children. In this process, the parents or surrogates involved, even when the latter are not of the animal's species, impart their own traits to the adopted young. For example, a young rhesus raised from birth in the investigator's home learns to respond sensitively and adequately to human language and action, but never acquires some of the patterns (e.g., a fear of snakes) supposedly "innate" in rhesus monkeys raised by their own mothers.

Early Deprivations

In contrast, young animals subjected to periods of solitary confinement, even though otherwise physically well cared for, do not develop normal initiative, physical stamina, or social relationships.

Character Deviance

Unusual experiences for the young animal may engram peculiar character-istics which persist through adulthood. For example, if a young animal is taught to work a switch and thus subject itself to increasingly intense but tolerable electric shocks as a necessary preliminary to securing food, for the rest of its life it may continue to seek such shocks even in the absence of any other immediate reward and may thus appear to be inexplicably "masochistic" to an observer un-acquainted with its special early history.

Early Brain Injuries

A remarkable finding was that adequate care and training in early life could in large part compensate for extensive brain damage in the newborn. Monkeys subjected to the removal of both temporal or parietal cortices at birth but given a protective and stimulating home environment thereafter suffered minor kinesthetic and affective impairments which could be revealed by special tests or during periods of sensory deprivation, but developed otherwise normal and ade-quate individual and social adaptations. On the other hand, in the absence of such special care and training the effects of brain damage, including bilateral lesions in the thalami, amygdalae, and in cerebral areas 13, 23, and 24 were much more devastating in the young than in adult animals, and did not amelio-rate induced experimentally neurotic behavior as effectively as in the case of adults.

Early "Psychological" Trauma

These were even more devastating: If the young animal was subjected to un-predictable or exceedingly severe conflicts between counterposed desires and aversions, or even mutually exclusive satisfactions, it developed deeply ingrained inhibitions, fears, rituals, somatic disorders, social maladjustments and other aberrations of behavior which became highly elaborate and more difficult to treat than those originating in adulthood.

Social Relationships

Group Organization

Animal societies in the laboratory as well as in the feral state organize them-selves in hierarchies of relatively dominant and submissive members, with leadership and privilege generally preempted, not by size or strength alone, but in accordance with special aptitudes and "personality" skills. However, these relationships could be modified in the following significant ways.

PARASITISM AND TECHNOLOGICAL SOLUTIONS

Cooperation

Under special experimental conditions, a cat or a monkey could be trained to operate a mechanism that produced food for its partner, who then reciprocated in "mutual service." In most pairs, however, this pattern soon devolved into either (a) a situation in which neither animal would work for the other so that both starved; or (b) a fairly stable "industrial" relationship in which one animal (the "worker") operated the feeder mechanism sufficiently frequently for both while receiving only part of the profits. However, two such workers were sufficiently "intelligent" (i.e., possessed of unusually high perceptive-manipulative capacities) to jam the feeding mechanism so that it operated continuously, thus solving the social problem by a form of automation.

"Altruism"

Some macaques continued to starve for days rather than pull a lever to secure readily available food if they had learned that this also subjected another macaque to an electric shock. Such "succoring" behavior was apparently less dependent upon the relative age, size, or sex of the two animals, than on (a) their individual "character" and (b) whether or not they had been mutually well-adjusted cagemates for a period of 3-6 months.

Aggression

Conversely, aggression in the sense of actual fighting between members of the same species to establish various relationships was minimal; primacy and dexterity manifested by only occasional gestures of preemption were nearly always sufficient to establish dominance and privileges. Indeed, physical combat appeared only under the following special circumstances.

(a) When an animal accustomed to a high position in its own group was transferred to one in which it came into direct conflict with new rivals previously accustomed to dominance.

(b) When a female with increased status derived from mating with a dominant male attacked others in the group.

(c) When a dominant animal was subjected to a rebellion or a conjoint territorial defense by an alliance of subdominants.

(d) When a dominant animal, by being made experimentally neurotic as described below, fell to a low position in its own group, and thereafter expressed its frustrations by physical attacks both on inanimate and living objects in its environment.

EXPERIMENTAL "NEUROSES" AND "PSYCHOSES"

Methods of Induction

Proceeding from the control observations described in the preceding section, we determined that, in accordance with the fourth biodynamic principle, marked and persistent deviations of behavior could be induced by stressing the animal between mutually incompatible patterns of survival: as, for instance, requiring a monkey to secure food after a conditional signal from a box which might unexpectedly contain a toy snake—an object as symbolically dangerous to the monkey as a live one, harmless or not, would be. In this connection, we have further amended Freudian doctrine by demonstrating that "fear" in the sense of dread of injury need not be involved at all; i.e., equally serious and lasting neurotigenic effects can be induced by facing the animal with difficult choices among mutually exclusive satisfactions—situations that parallel the disruptive effects of prolonged uncertainties and consequent anxieties and ambivalences in human affairs.

Either form of conflict induced physiologic and mimetic manifestations of anxiety, spreading inhibitions, generalizing phobias, stereotyped rituals, "psychosomatic" dysfunctions, impaired social interactions, and other persistent regressions and deviations of conduct.

Constitutional Influences

Animals closest to man showed symptoms most nearly resembling those in human neuroses and psychoses, but in each case the "neurotic syndrome" induced depended less on the nature of the conflict (which could be held constant) than on the constitutional predisposition of the animal. For example, under similar stresses, spider monkeys reverted to infantile dependencies or catatonic immobility, cebus developed various "psychosomatic" disturbances including functional paralysis, whereas` vervets became diffusely aggressive, persisted in bizarre sexual patterns, or preferred hallucinatory satisfactions such as chewing and swallowing purely imaginary meals while avoiding real food to the point of self-starvation.

Methods of Therapy

Since we induced experimental neuroses in animals not only to study their causes and variations, but primarily to search for the principles of therapy, this portion of our work was assigned the most time and effort. After the trial of scores of procedures only nine general techniques, significantly parallel to those used with human patients, proved to be most effective in ameliorating neurotic or psychotic symptoms in animals. In briefest statement, these methods were:

1. *Satisfaction* of one of the conflictful biologic needs, such as frustrated hunger, thirst or sex.

2. *Removal* from the laboratory to a less stressful environment; by analogy to humans, a temporary vacation, a better home, job, or climate.

3. *Forcing the solution* of a motivational impasse by directed stress, but carefully keeping them within the organism's tolerance; e.g., conducting an acrophobic by imposed stages to increasing heights.

4. *Furnishing opportunities* for the utilization of acquired skills for reasserting mastery over the environment; in clinical parallel, inducing a crashed but unhurt fighter pilot to recover his threatened feeling of mastery by flying another plane immediately.

5. *Providing "social examples" in a group of well-adapted ("normal") organisms:* i.e., sending a "problem child" to a "good" school where children behave in the desired manner.

6. *Retraining* the animal by individualized care and guidance, the experimenter acting as a "personal therapist" conversant with the nature of childhood or later experiential conflicts and their neurotigenic effects; or, as in psychoanalysis, helping the patient recall and reexperience conflicts and dispelling their current residues by "corrective emotional experience" vis-a-vis the therapist and elsewhere in his adult world.

7. *Utilizing electroshock* or inducing other forms of cerebral anoxia and diaschisis to disrupt undesirable patterns of behavior.

8. *Performing various brain operations* to induce similar neuropsychologic disorganization; in monkeys, lesions in regions corresponding to cortical areas 12, 23, and 24 were most suitable for this purpose. Another highly significant finding was that the aftereffects of any brain operation depended not only on its site and extent but also on the preceding experiences and characteristics of the animal. For example, bilateral lesions of the dorsomedial thalamic nucleus im-

paired the learning capacities of a normal animal but left it gentle and tractable; in contrast, the same operation rendered a neurotic animal irritable and vicious.

9. *Finally, the use of various drugs* for the temporary amelioration of disturbing perceptions and conflictual reactions, thus facilitating other methods of therapy. Additional pharmacologic observations were as follows.

(a) *Effectiveness.* Our data indicated that most of the recently promoted "tranquillizers" (meprobamates and phenothiazines) were less generally effective than drugs long tested in clinical therapy; e.g., alcohol, paraldehyde, and the barbiturates and bromides.

(b) *Preventive action.* Such mildly obtunding drugs, if administered before subjecting an animal to stress or conflict, would also partially prevent the after-effect of an otherwise traumatic experience; so also, we are apt to take a "bracer" before asking for a raise, proposing marriage, or contemplating some other supposedly hazardous undertaking.

(c) *Addiction.* However, a neurotic animal permitted to experience the relief from fear and inhibition produced by alcohol would tend to prefer alcohol to nonalcoholic food and drink and thus develop a dipsomania which would persist until its underlying neurosis was relieved by other means. In humans similar psychopharmacologic effects of addictive drugs are demonstrably operative but are, of course, complicated by ethnic, cultural, and highly individualized symbolic vectors.

BEYOND THE ABOVE. . .

These experiments, then, in context with those of Sechenov, Pavlov, Gantt, Delgado, Skinner, Mirsky, Brady, and many others cited in my previous reviews of the literature[2] have helped clarify some basic issues of etiology and therapy. However, we must concede that mice in mazes, cats in cages, and monkeys in pharmacologic hazes cannot explain all human history, since man differs from all other animals in having developed three ultimate (Ur) axioms, beliefs, assumptions, aspirations, illusions—call them what you will—that have intimately imbued his behavior. I have dealt with these strange but ubiquitous determinants at considerable length elsewhere[3]; here I can only briefly frame them in context as follows.

First, an urgent assertion of omnipotence, invulnerability and immortality that leads man to seek absolute control over his material universe and deny death itself.

---

[2]Cf. particularly my chapter in "The Biodynamic Roots of Psychoanalysis" in J. Marmor (Ed.), *Frontiers of Psychoanalysis.* New York: Basic Books, 1968.
[3]Cf. my text *Practice of Dynamic Psychiatry.* Philadelphia: W. B. Saunders Co., 1956.

Second, a persistent need for human relationships that, despite destructive debacles of distrust, have led from the mother—child through the family, clan, tribe, state, and nation to a growing imperative toward the brotherhood of man.

And finally, an existential faith that man's being has an enduring dignity and significance in some universal philosophic, theologic, and essentially eschatologic system that extends beyond the here and now into eternity.

In essence, though no human being can ever be *completely* certain of his health, friends, or philosophy, adequate modicums of security in each of these spheres are essential to his welfare, and all methods of medicopsychiatric therapy are effective only insofar as they restore physical well-being, foster more amicable interpersonal relationships, and help the patient amend his beliefs so as to render them more generally acceptable and useful without impugning his individuality, infringing on his essential freedom, or stereotyping his intellect and imagery. Indeed, this constitutes the "art" of comprehensive therapy applied seriatim as follows.

To begin with, if patients are to continue to come to us at all confident that we shall furnish the care they seek, we must regain, cherish, and increase public regard for us not only as skilled technicians, but also as dedicated humanitarians deserving the highest respect and confidence. Differences of professional opinion, as in any scientific field, are acceptable, but public polemics, a trade-union facade, and blatant economic and political partisanships diminish our stature and impair the trust we need if we are to serve our patients to their and our best advantage. No one of us can alone undertake this essential restoration of our former prestige and influence; it is a task—and an important one— for all of us.

Next, in our direct handling of individual patients, we must discard our defensive armour of aloof "professional dignity," and accept each supplicant not as a diagnostic "challenge," or a recipient of specific therapy for his "psychopathology" (a repulsive solecism), and least of all as only another research datum, but as a *troubled human being seeking comfort and guidance* as well as mere relief from "physical" or "mental" suffering. These larger requirements can be met as follows.

Regardless of whether the patient's complaints are considered as primarily "organic" or "neurotic," bodily discomfort and dysfunction should be ameliorated by every medical and surgical means available including, when indicated, conservatively prescribed and frequently changed sedatives and hypnotics temporarily useful to dull painful memories, diminish apprehension, and quiet agitation. However, we recognize that we must strive concurrently to reevoke the patient's initiative, restore his lost skills, and encourage him to regain the competence and respect that can come only from useful accomplishment (Uradaptation I).

But since no man is an "Island unto himself," the patient may be deeply concerned about financial, sexual, marital, career, and related problems that may also seriously affect his social well-being. This involves an exploration, varying in depth and duration but always discerning and tactful, of the patients's characterologic assets and vulnerabilities, the attitudes, values, and incentives he derived from his past experiences, the channelling effects of past successes or the disruptive residues of past and current tribulations, the "conscious" or "unconscious" residues (the distinction is not always germane) they have left in his current patterns of effective (normal), socially ineffective (neurotic), or bizarrely unrealistic (psychotic) conduct, the ways in which these patterns relieve or exacerbate his current difficulties, and which of them are most accessible to various methods of treatment.

In general, the transactional aspects will consist of using anamnestic review and reconstruction, gentle reasoning, personal example, and progressive reorientative experiences to help the patient modify his prejudices, correct past misconceptions, abandon infantile or childlike patterns of behavior that have long since lost their effectiveness, revise his values and objectives, and adopt a more realistic, productive, and lastingly rewarding ("mature") style of life. In this skillfully guided reeducation (good psychotherapy is about as "nondirective" as good surgery), the enlightened cooperation of his family, friends, employer, or others may, with the patient's consent, be secured and utilized to the full. By such means the patient's second Ur-adaptations will be strengthened by renewed communal solidarity and security—a *sine qua non* of comprehensive treatment.

Lastly, to mitigate the third, or cosmic Ur-anxiety, the patient's religious, philosophic, or other convictions, instead of being deprecated and undermined, should be respected and strengthened insofar as they furnish him with what each of us requires: a belief in life's purpose, meaning, and value. In this deepest sense, medicine, and its subdiscipline, psychiatry, being humanitarian sciences, can never be in conflict with philosophy or religion, since all seem to be designed by a beneficent providence to preserve, cheer, and comfort man and thereby constitute a trinity to be respected by any physician deeply concerned with man's health and sanity. Indeed, with respect to these latter terms, it is of historic–philologic significance that *sanatos* implied to the ancients indissolubility of physical and mental functions (*mens sanis in corpora sano*) and that the word *health* can be traced to the Anglo-Saxon *hal* or *hōl*, from which are derived not only physical *haleness* and *healing*, but the greeting "Hail, friend!" and the concepts of *wholeness* and *holiness*. Ergo, once again, Greek, Roman, and Gaul have bequeathed to us, in the rich heritage of a syncretic language in which "reality" and "illusion" merge, their penetrating recognition of the indissoluble physical, social, and philosophic components of health and sanity.

BIBLIOGRAPHY

Relevant Publications by the Author

Books

Practice of Dynamic Psychiatry. Philadelphia: W. B. Saunders, 1955.
Principles of Dynamic Psychiatry. 2nd Ed. Philadelphia: W. B. Saunders, 1961.
(Ed.): *Psychiatry: East and West.* New York: Grune & Stratton, 1968.
Theories and Therapies in Dynamic Psychiatry. New York: Jason Aronson, 1973.

Articles

Anxiety and the art of healing. In J. H. Masserman (Ed.), *Current psychiatric therapies* (Vol. I). New York: Grune & Stratton, 1961, pp. 216–238.
Or shall we all commit suicide? In J. H. Masserman (Ed.), *Current psychiatric therapies* (Vol. II). New York: Grune & Stratton, 1962, p. 273.
Contemporary psychotherapies: a review. *Archives of General Psychiatry*, 1962, *8*, 461–464.
Humanitarian psychiatry. Bulletin of the New York Academy of Medicine, 1963 *39*, 533–544.
Anxiety: the protean source of communication. In J. H. Masserman (Ed.), *Science and psychoanalysis* (Vol. VIII). New York: Grune & Stratton, 1965.
Man's eternal anxieties and compensating illusions. Camp Memorial Lecture, Part I. *Illinois Medical Journal*, 1965, *127*, 375; Part II, *ibid.* 1965, *127*, 561.
The timeless therapeutic trinity. In J. H. Masserman (Ed.), *Current psychiatric therapies* (Vol. VII). New York: Grune & Stratton, 1967.
Science, psychiatry and human values. *Proceedings of Fourth World Congress of Psychiatry, Psychotherapy and Psychosomatics* 1967, *1*, 703–707.
Asymptotic systems of survival. In W. Gray, F. V. Duhl, and N. D. Rizzo (Eds.), *General systems theory.* Boston: Little, Brown, 1969.

Animal Models of Violence and Hyperkinesis
Interaction of Psychopharmacologic and Psychosocial Therapy in Behavior Modification

SAMUEL A. CORSON, E. O'LEARY CORSON, L. EUGENE ARNOLD, AND WALTER KNOPP

WHY ANIMAL MODELS OF PSYCHOPATHOLOGY?

In reviewing the dramatic discrepancy between the many favorable clinical reports on meprobamate and a number of negative laboratory studies on this drug, Gerard (1957), in his usual elegant and succinct manner, summarized the situation as follows: "On the one hand, laboratory studies indicate that meprobamate is quite inert; on the other hand, a great number of takers seem to experience some benefit, and clinical reports included in these pages—many seemingly well controlled and convincing enough as reported—indicate a definite action. From this, one must conclude either that the experimenters have not yet

SAMUEL A. CORSON, E. O'LEARY CORSON, L. EUGENE ARNOLD, and WALTER KNOPP • Laboratory of Cerebrovisceral Physiology, Division of Behavioral and Neurobiological Sciences, Department of Psychiatry, College of Medicine, Ohio State University, Columbus, Ohio.

111

found the right thing to test—which would not be surprising, since we are dealing with agents active on the nuances of complex human behavior for which it is difficult to find electrical or chemical indicators, either in the laboratory or in the patient—or else the clinical impressions are wrong. . . . I doubt that the latter is the case, especially because of the genuine awareness of the problem of controls that exists today among the better clinicians and laboratory workers handling these problems."

Our studies on the effects of meprobamate on dog models of anxiety (Corson, 1966a and 1966b; Corson & Corson, 1967; Corson, 1969) have demonstrated that Gerard hit the nail on the head. The laboratory workers indeed failed to test the drug on clinically relevant parameters.

We suspect that the main reason why the laboratory workers found meprobamate to be inert was that they were testing the drug on normal animals. In our studies on normal dogs, meprobamate, in therapeutic doses, had no measurable influence on behavioral or physiologic variables, such as heart rate, respiration, or renal function. However, in dogs with experimental anxiety, meprobamate exhibited remarkably normalizing effects by *inhibiting conditional visceral responses to an aversive environment, without significantly affecting conditional classical motor defense reflexes or conditional discriminated avoidance or escape responses.* The laboratory workers who reported that meprobamate was inert were in error not only because they studied the effects of this drug on normal animals, but also because they recorded only conditional avoidance or escape responses.

We previously reported (Corson, Corson, & England, 1963) the case of a dog which exhibited a highly selective aggressive behavior toward one kennel attendant who had apparently abused the dog before this animal was acquired by our laboratory. When this dog was on a chronic meprobamate dosage that inhibited all conditional visceral responses to an aversive environment, the aggressive attacks on the attendant were just as easily provoked and as well directed as they had been before the dog was treated with meprobamate. It is this selective action on exaggerated conditional visceral reactions that endows anxiolytic drugs with their useful therapeutic properties. Such selective effects could be demonstrated only on animal models of psychopathology.

SEARCH FOR ANIMAL MODELS OF HYPERKINESIS

The problem of hyperkinesis in children and the so-called paradoxical calming effects of amphetamine have remained perplexing chiefly because of the lack of suitable animal models. The hyperkinetic syndrome (also called minimal brain dysfunction, hyperkinetic behavior disorder, minimal brain damage, organic behavior disorder, hyperkinetic impulse disorder, etc.) has been so well documented in the literature (e.g., Bradley, 1950; Laufer & Denhoff, 1957; Millichap, 1968; Kornetsky, 1970) that Jenkins (1970, p. 681) was able to state that "few clinical pictures in child psychiatry are so widely recognized and accepted as this one." Because of the plethora of literature, no attempt is made here at exhaustive referencing. Only representative sources are cited.

The syndrome, as described by the above-mentioned authors, includes overactivity, restlessness, distractibility, short attention span, misbehavior, impulsivity, temper tantrums, unpredictability, underachievement, poor ability to learn from experience, poor conditionability, and low frustration tolerance. Some hyperkinetic children also exhibit poor coordination, late developmental milestones, and specific learning disability. Though not all hyperkinetic children show all elements of the syndrome, approximately 4% of schoolchildren show enough of these symptoms to permit a diagnosis (Millichap, 1968). Masland (1965) estimated the incidence of the hyperkinetic disorder to be as high as 5–10% of the school population.

This constitutes a public health and mental health problem of immense magnitude (Eisenberg, 1957). Without treatment, many of these children develop secondary emotional problems severe enough to outlast the primary syndrome and to persist into adult life. Many of these children, though of normal intelligence, cannot function in school and may drop out, sometimes turning to delinquency. This not only constitutes a waste of human resources, but also aggravates the problems of law enforcement, mental health, welfare, and other public services.

About half of such children have abnormal EEGs, usually of a diffuse nature. Whether the other hyperkinetic children are truly "brain damaged," partly depends on whether one accepts such "soft neurological signs" as hyperkinesis, motor overflow, clumsiness, strabismus, and articulatory speech defects. However, there has been a tendency recently to accept some sort of neurological problem as the basis for the hyperkinetic syndrome (Nichamin & Barahal, 1968), at least in the sense of "maturational lag" or immaturity of the nervous system (Jenkins, 1970).

Unfortunately, the exact nature of the neurological deficit is not known. Several hypotheses are currently in vogue: some (e.g., Laufer, Denhoff, & Solomons, 1957; Laufer & Denhoff, 1957; Laufer, 1962) believe that the basic problem is in the diencephalon. On the basis of their studies with the photo-Metrazol test, Laufer et al. suggested that the decrease in inhibitory influences from the diencephalon "exposes the cortex to unusual storms." Another hypothesis is that the cortex does not adequately inhibit brain-stem impulses (Freedman & Kaplan, 1967). According to this hypothesis, hyperkinesia is due to a "reverberating feedback type of circuit between the cortex and the diencephalon" because of an assumed discrepancy in the functional maturation between the cortex and the diencephalon. Snyder, Taylor, Coyle, & Meyerhoff (1970) have published studies raising the question whether one of the dopaminergic systems may be the problem site. Wender (1971) has found in some hyperkinetic children abnormally low platelet serotonin levels suggesting a metabolic error.

The reason for these divergent hypotheses may well be that there is more than one possible cause for the syndrome. The lack of clarity about etiology and mechanisms of symptom production has made it difficult to develop a scientific approach to treatment. The current status of this problem was elegantly summarized by Wender (1971).

Wender pointed out that hyperactivity is not invariably a component of this disabling syndrome, since some of these children may actually be *hypoactive* (and become normalized by central nervous system stimulants). Wender, therefore, prefers to use the term "Minimal Brain Dysfunction" (MBD). Wender characterizes the MBD child as exhibiting: (a) a diminished susceptibility to pleasure and pain which is paralleled by a decreased susceptibility to positive and negative reinforcement and consequent impaired conditionability; (b) poorly modulated (often excessive) level of activity. Wender hypothesized that: (1) MBD children have an abnormality in the metabolism of monoamines (norepinephrine, dopamine, or serotonin); (2) this biochemical abnormality affects behavior by the impairment produced in: (a) the reward mechanism of the brain, and (b) the activating system of the brain.

A candid statement on the current "state of the art" of the MBD syndrome was given by Daniel X. Freedman (1971) in his capacity as chairman of a panel "On the Use of Stimulant Drugs in the Treatment of Behaviorally Disturbed Young School Children." The statement reads:

> We know very little about definite causes. The disorders have been ascribed to biological, psychological, social or environmental factors, or a combination of these. . . . The neurological and psychological control of attention is an important but incompletely researched topic, as are the nutritional, perinatal, and developmental factors. Thus, *in many instances, it is not yet possible even to speculate as to original causes* [emphasis ours].

Some of the MBD children respond favorably to CNS stimulants like amphetamine and methylphenidate (Bradley, 1937, 1950, 1953; Burks, 1964; Clements & Peters, 1962; Conners, Eisenberg, & Barcai, 1967; Conners & Eisenberg, 1963; Conrad, & Insel 1967; Creager & Van Riper, 1967; Eisenberg, 1957, 1962, 1966; Eisenberg, Lachman, Molling, Lockner, Mizelle, & Conners, 1963; Eisenberg, Conners & Sharpe, 1965; Epstein, Lasagna, Conners, & Rodriguez, 1968; Knights & Hinton, 1969; Laufer, 1962; Laufer et al., 1957; Wender, 1971; Arnold, Wender, McCloskey, & Snyder, 1972). The percentage of CNS drug responders among MBD children varies (depending on the author) from 30 to 70%.

The truth of the matter is, that the MBD syndrome, far from being a homogeneous clinical entity, has in fact become a diagnostic wastebasket with nebulous blurred definitions which make it impossible to compare studies by different clinicians, as was pointed out by Arnold, Kirilcuk, Corson & Corson (1973), Fish (1971), and Wender (1971). Among the MBD children who are helped by amphetamine, there is a wide variation in dosages required for normalization. Moreover, some MBD children respond favorably to amphetamine and not to methylphenidate and vice versa. Some MBD children respond well to d-amphetamine, whereas others may do better with a racemic mixture of this drug (Bradley, 1950; Laufer & Denhoff, 1957). Some MBD children are actually made worse by CNS stimulants, while others may show improvement with a neurochemically opposite form of therapy, e.g., chlorpromazine or haloperidol (K. K. S. Voeller and John Kempf, personal communication).

According to H. R. Huessy (1967 and personal communication), "Children with behavior disorders are extremely variable in their responses to medication. . . . We have been totally unable to find any predictive factors so far. . . . For years I have been saying that our research in behavior disorders is at a dead end until we can find an experimental animal." This presentation deals with descriptions of such animal models developed in our laboratory.

BACKGROUND STUDIES WHICH LED TO THE DEVELOPMENT OF ANIMAL MODELS OF HYPERKINESIS AND VIOLENCE

The development of these animal models stems from longitudinal studies on psychophysiologic individuality conducted on several breeds of dogs in our laboratory during the past decade. Our aim has been to delineate individual differences in integrated adaptive physiologic and behavioral responses of dogs exposed to controlled repeated psychologic stress, as a basis for elucidating mechanisms for resistance or susceptibility of individuals with different constitutional makeup to the development of psychosomatic or behavioral disturbances.

We utilized Pavlovian conditioning techniques with unavoidable electro-cutaneous reinforcement for the production of psychologic stress. Under these conditions some of the dogs developed an almost inextinguishable quintet of reactions exhibited not only to conditional or unconditional stimuli but also to the entire conditioning-chamber complex. This quintet of responses consisted of sustained tachycardia, persistent hyperpnea, excessive salivation, a vasopressin-type antidiuresis, and increased energy metabolism (Corson, 1966a and 1966b; Corson, Corson, & Kirilcuk, 1970; Corson, 1971; Corson & Corson, 1971; Corson, Corson, Kirilcuk, Knopp, & Arnold, 1972a; Corson, Corson, Kirilcuk,

Arnold, & Knopp, 1972b). Other dogs exposed to the same kind of psychologic stress failed to exhibit this quintet of stress responses or exhibited only temporarily some of the components of the quintet and soon showed physiologic adaptation to the stressful environment.

In our previous publications we referred to the dogs exhibiting the sustained quintet of reactions as *antidiuretic dogs*, whereas the rapidly adapting dogs we called *diuretic dogs.* This terminology was used chiefly because this happened to be the first parameter which we recorded in these animals and because the renal response appeared to be the most persistent characteristic of these two types of dogs.

Subsequent experiments demonstrated that the antidiuretic dogs, when exposed to unavoidable stress (Pavlovian conditioning with electrocutaneous reinforcement), responded physiologically to the entire conditioning-room complex as though they were engaged in intense muscular activity. The visceral reactions were reminiscent of those described by Cannon (1932) as the "fight or flight" reaction. Measurements of oxygen consumption and CO_2 production indicated that the primary response of these dogs to an aversive environment is increased heat production (Corson, 1971). This in turn stimulates the hypothalamic thermoregulatory center to increase heat-dissipation mechanisms: polypnea and salivation. The antidiuresis thus serves the function of conserving body water, so that it may become available for thermoregulatory salivation.

The diuretic dogs do not exhibit the persistent quintet of emotional turmoil because their energy metabolism in the aversive environment is not increased. Therefore, they have no need for invoking increased heat-dissipation mechanisms.

The antidiuretic dogs represent an experimental model of psychosomatic disturbances engendered by emotional stress. It is highly instructive that when antidiuretic dogs are exposed to similar conditional and unconditional stimuli in another room where they are permitted to gain control of the environment by developing conditional avoidance responses, the components of the quintet of emotional reactions become markedly attenuated (Corson, 1969, 1971).

Our initial studies were conducted on mongrel dogs. Later we shifted to the investigation of several distinct breeds, including dogs obtained from Dr. John Paul Scott's colony, originally at the Jackson Laboratory and later at Bowling Green State University. Thus far we have studied about 70 dogs.

The rationale for choosing dogs as our experimental animals is based on the fact that a great variety of standard breeds is available. As pointed out by Scott and Fuller (1965)

> We chose the dog because it shows one of the basic hereditary characteristics of human behavior: a high degree of individual variability. . . . The dog is a veritable genetic gold mine. Besides the enormous differences between breeds, all sorts of individual differences appear at the shake of a genetic pickaxe, in this case the technique of mating two closely related animals. Anyone who

wishes to understand a human behavior trait or hereditary disease can usually find the corresponding condition in dogs.

Another reason for choosing dogs is the fact that these animals exhibit a variety and intensity of emotional reactions approximating those shown by human beings. These animals are large enough to permit biochemical and endocrine studies of body fluids. They are generally cooperative and relatively easily trained and they will tolerate reasonably well surgical implants and experimental paraphernalia without having to resort to such extreme unphysiological and stressful restraints as a primate restraining chair. Moreover, a wealth of data on higher nervous functions has been accumulated by the Pavlovian school. Stockard, Anderson, and James (1941) presented a good amount of material on genetics in relation to endocrine morphology and some behavioral parameters. And finally, we have the elegant studies of Scott and Fuller (1965) on genetics and the social behavior of dogs.

In the course of our attempts to characterize the psychobiologic nature of different types of dogs, we encountered a group of dogs which could not be trained at all in a Pavlovian stand, even before the introduction of any conditional or unconditional nociceptive stimuli. These dogs fought violently against the restraining Pavlovian-type harness and tried to bite and chew everything within reach. Neither time nor patience was of any avail; in fact, many of these dogs became less manageable in repeated experimental sessions. Similarly, positive or negative reinforcement procedures proved to be useless. Antipsychotic drugs (e.g., chlorpromazine) or antianxiety substances (e.g., meprobamate) did not help, even when used in high dosages.

Because these dogs could not be conditioned, our technicians referred to them as "stupid," just as teachers often refer to hyperkinetic children as being stupid. One had the impression that these animals could not filter information and could not narrow and focus their attention. This, together with failure of these animals to respond to positive or negative (aversive) reinforcement, suggested to us that we may be dealing with a syndrome comparable to childhood hyperkinesis. It then occurred to us that perhaps these dogs might respond favorably to amphetamines in a manner similar to the responses of some hyperkinetic children to central nervous system stimulants.

We therefore decided to investigate the effects of amphetamines on several hyperkinetic and on normal well-behaved dogs in whom we previously developed stable Pavlovian conditional motor defense responses. This presentation will describe results of our observations on the beneficial effects of amphetamines on some hyperkinetic dogs as well as some unfavorable influences of these drugs on normal dogs. We shall also present some preliminary observations on attenuating effects of amphetamines on emotionally induced visceral turmoil in an antidiuretic dog.

METHODS

More than 900 experiments were performed on 18 dogs, 8 of which were hyperkinetic. Our dogs were kept in air-conditioned runs (9 × 3 ft) which included doghouses where dogs could sleep or rest. In addition, the dogs were periodically taken out for walks. This type of dog maintenance permitted a reasonably normal physiologic and psychologic status of the animals and helped to maintain good sanitation conditions and clean dogs. Maintaining dogs in cages is physiologically and psychologically unsound. The dogs were given standardized food rations designed to maintain a constant normal weight in each dog.

In the classical conditioning experiments, the dogs were brought to the conditioning room and mildly restrained on a Pavlovian stand. After the attachment of electrodes and other instrumentation, the dogs were left in the conditioning chamber for two hours. In addition to the data obtained from observation of their general behavior, the following parameters were recorded: EKG, heart rate, respiration, EMG from surface back muscles, urine flow, leg lift (conditional and unconditional motor defense responses, using electrocutaneous stimulation reinforcement), rectal and skin temperatures, and conditional cardiac and respiratory responses. In some experiments oxygen consumption was recorded continuously by means of a modified Beckman Biomonitor. This technique will be published elsewhere. Urinary catecholamines and electrolytes were analyzed by means of a Technicon Autoanalyzer.

Pavlovian motor defense and operant discriminated avoidance conditional responses were developed by electrocutaneous reinforcement of auditory stimuli.

Baseline control data were obtained in the well-behaved group of 10 dogs; these data were highly reproducible. Numerous attempts were made to secure baseline control data in the untrainable group of 8 dogs, sometimes as many as 30 experiments on a single animal. However, it was usually impossible to keep these animals on the stand for more than 15–20 min because they would bite through and destroy the harness and electrode attachments and would also tend to hurt themselves. We therefore began to study the possible calming effects of the following drugs, administered per os one hour before the experiment: chlorpromazine, 3–14 mg/kg/24 hours; meprobamate, 40–120 mg/kg/24 hours; and *d*- and *l*-amphetamine, 0.25–4.0 mg/kg. The drug tablets were administered by concealing them in a small ball of chopped meat which was offered to the dog after a standard period of withholding food and water. In control experiments, the dogs were offered the same amount of chopped meat (without drug) one hour before the experiment.

RESULTS

The Influence of *d*- and *l*-Amphetamines on Naturally Occurring Untrainable Hyperkinetic Dogs

Thus far we have studied the effects of amphetamines on eight hyperkinetic dogs:

1. Jackson (cocker spaniel X beagle hybrid, male, whelped 69-02-01; age 1.7 yrs).
2. Simone (beagle X telomian hybrid, J. P. Scott #21, female, whelped 68-10-04; age 2 yrs).
3. Dunco (beagle X telomian hybrid, J. P. Scott #17, male, littermate of Simone; age 2 yrs).
4. Chester (English shepherd R9–2145-5129, male, whelped 59-05-11; age 11 yrs).
5. Whiskey (husky X German shepherd hybrid, male, whelped 67-04-01; age 4 yrs).
6. Perri Dot (wirehair fox terrier R332452, female, whelped 62-07-10; age 8 yrs).
7. Jule (wirehair fox terrier R319269, male, whelped 62-03-10; age 8 yrs).
8. Saumie (cocker spaniel, male, age 7 yrs

The ages of the dogs are given in relation to the first entrance of the animal into the experimental series.

We also studied the influence of *d*- and *l*-amphetamines on two hyperkinetic cats, one of which was also violent.

As mentioned above, the hyperkinetic behavior of these dogs could not be ameliorated by either negative or positive reinforcement procedures or by chlorpromazine or meprobamate.

Of the eight hyperkinetic dogs we studied, the first five dogs listed above exhibited dramatic normalizing responses to both *d*- and *l*-amphetamine. The calming effect appeared about one hour after the very first oral administration of the drug. There were marked and stable individual differences among the five dogs in the optimal normalizing dosages of the amphetamines, the oral dosages for the *d*-isomer varying from 0.2 mg/kg to 1.0 mg/kg.

Dog number 8 (Saumie) was a patient in our veterinary clinic. This dog was brought to the clinic because of his unpredictable violent and vicious behavior. The results on this animal are described below (p. 122).

Preliminary comparative studies on the effectiveness of the two amphetamine isomers were made on three hyperkinetic dogs (numbers 1–3). In general, it was necessary to use about 3–4 times as much *l*-amphetamine as the *d*-isomer in order to achieve the same calming effect.

Amphetamines exhibited no calming effects on dogs number 6 and 7. If anything, we had the impression that these dogs became even more unmanageable in the Pavlovian stand under the influence of these drugs.

The behavior of dog number 7 (Jule) was not typically hyperkinetic. He seemed to tolerate the Pavlovian conditioning restraints, but he could not learn to discriminate between the excitatory and inhibitory conditional stimuli, reacting violently to both tones. Thus far we could not discover any form of useful therapy for this learning disorder.

The same dosages of amphetamines which calmed hyperkinetic dogs produced marked restlessness and agitation in normal dogs with extensive experiences in Pavlovian conditioning paradigms. While under the influence of these drugs, the dogs lost their ability to discriminate between excitatory and inhibitory conditional stimuli; they overreacted to both types of stimuli and exhibited inappropriate intersignal responses.

The Influences of Amphetamines on Learning in Hyperkinetic Dogs

As mentioned previously, while the dogs were hyperkinetic, it was impossible to determine whether these animals could develop conditional responses. Regardless of the stimuli presented, these dogs continued to fight violently against the restraints. Under the influence of amphetamines, the dogs tolerated well the essential arrangements required for conditioning experiments. In fact, after the very first dose of the drug, the dogs behaved on the Pavlovian stand as though they had been trained for many months, suggesting that amphetamines may produce these normalizing effects by facilitating the retrieval of information which may have been stored in the central nervous system.

Conditioning experiments were then performed in three of the hyperkinetic dogs. Pavlovian conditional motor defense reflexes were developed in dogs Jackson and Dunco. In dog Simone (a littermate of Dunco) discriminated conditional avoidance responses were established.

Under the influence of amphetamine, all three dogs learned very rapidly, thus indicating that these dogs were Pavlovian or operant "school dropouts" not because of "canine mental retardation." Just as in the case of some hyperkinetic children, the learning disability of these dogs was a secondary effect resulting from their nonadaptive, non-goal-directed hyperactivity.

In dog Simone, *d*-amphetamine (0.2 mg/kg) was used for only five days. Avoidance responses appeared on the third day of the drug regimen. After the five-day amphetamine series, the experiments were continued without drugs, the dog reaching a criterion of 100% avoidance after 8 experimental no-drug days. Differentiating (inhibitory) tones were introduced on the fourth no-drug day.

After four more experimental sessions, this dog developed good discrimination, reaching a criterion of 90–100%.

In the case of the Pavlovian experiments, the two dogs had to be maintained on d-amphetamine throughout the entire period of conditioning until the animals developed stable differentiated conditional motor defense, cardiac, and respiratory responses. In Jackson the optimal dose of d-amphetamine was 1.0 mg/kg; in Dunco, it was 0.2 mg/kg. *The learned responses remained after the drug was discontinued*, although some of the hyperkinetic behavior still persisted in a no-drug situation, particularly in dog Jackson. After 10–14 weeks of combined amphetamine and psychosocial therapy, the hyperkinetic behavior became significantly ameliorated even in no-drug situations. The good behavior persisted, for several weeks, as long as the dogs continued to be trained in the Pavlovian stand, although no drugs were administered.

However, when these dogs were left in the kennel for some 6 months without any further drug administration and without further experiences in conditioning experiments, a good deal of the hyperkinetic behavior returned when conditioning experiments were resumed. This is in contrast to the more lasting effects of amphetamine on violent behavior, as will be described later in this presentation.

The Inhibition of Violent Behavior by Amphetamines

One of our hyperkinetic dogs, Jackson, was also persistently and incurably vicious to other dogs and to humans, including the caretakers who provided his daily food. This dog was donated to our laboratory by a "rejecting master," since neither the owner nor any members of his family were able to develop friendly, peaceful relations with this animal. In our laboratory kennels, Jackson would bark, growl, snarl, and try to bite anyone approaching his run (see Figure 1), so that our laboratory personnel finally refused to have anything to do with him. Our heroic attempts to train him to accept the discipline of a Pavlovian conditioning stand turned out to be futile and costly, since he tore the harness and all the cables and electrodes within reach. Thus, Jackson was finally given up as a hopeless "Pavlovian-school dropout."

Chlorpromazine or meprobamate were not effective in controlling either the hyperkinesis or the violent behavior. As can be seen in Figure 2, an oral dose of d-amphetamine (1 mg/kg) within a period of one hour dramatically transformed the incorrigible, vicious, antisocial warrior into a peaceful, cooperative, lovable dog who not only permitted himself to be petted but appeared to enjoy the social amenities, as evidenced by his general appearance and by the kind of whimpering one sees in properly reared pets.

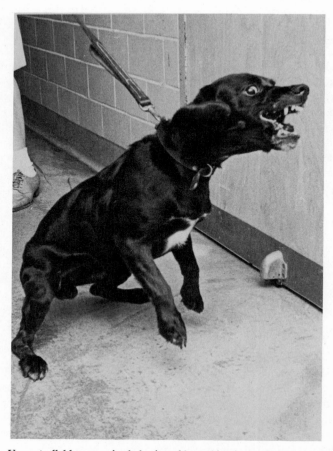

Figure 1. Uncontrollable aggressive behavior of hyperkinetic dog Jackson, male, 10.5 kg.

The pacifying effects of *d*-amphetamine lasted for about 5–7 hours. Administration of *l*-amphetamine was equally effective in the same dosage as the *d*-isomer; the only difference we observed was that the antiviolent effects of *l*-amphetamine lasted for only about 3–4 hours.

We also conducted preliminary investigations on another hyperkinetic dog (a patient in our veterinary clinic) who also exhibited uncontrollable violent behavior. The dog (Saumie, a cocker spaniel, male, 7 years old) appeared to be normalized in regard to hyperkinesis and violence by small doses of *d*-amphetamine (1 mg/kg). We have not yet tried lower dosages of *d*-amphetamine, nor do we have enough data on the effects of the *l*-isomer on this dog.

One of our two hyperkinetic cats also exhibited uncontrollable and unpredictable violent behavior. Amphetamines also normalized the behavior of these animals, although we do not yet have enough data regarding the comparative efficacy of the two isomers.

Because of lack of research funds, experiments on the dog Saumie and on the cats had to be discontinued.

In the case of dog Jackson, after a period of six weeks of amphetamine administration (1 mg/kg/day for a 5-day week), the drug was withdrawn without the reappearance of violent behavior. *At this writing, Jackson has been without any medication for over two years without the resurgence of his predrug aggressive traits.* It is still safe to offer Jackson a treat without the danger of having

Figure 2. Transformation of Jackson into a friendly, lovable, trainable dog one hour following oral administration of *d*-amphetamine (1 mg/kg).

one's hand bitten. In fact, this dog is so well behaved and lovable that one of our technicians wanted to take him home as a pet.

It should be mentioned here that Jackson's violent behavior may have a strong genetic component, as evidenced by the fact that the other four of his littermates exhibited similar aggressive traits. After we discovered the remarkable normalizing effects of amphetamine on Jackson, we tried to find the other littermates. It turned out that all these dogs were so badly behaved that they were eventually destroyed. Further support for the genetics hypothesis is furnished by the fact that Jackson has a cocker spaniel inheritance. We are told by dog breeders that they often encounter violent behavior in certain varieties of this breed of dogs. In our own experience, all the cocker spaniels we encountered in our laboratory were hyperkinetic and impossible to condition. Our other violent dog, which responded to amphetamine, Saumie, was also a cocker spaniel.

Not all of our violent dogs responded favorably to amphetamines. We have two wirehair fox terriers who would invariably try to fight violently with each other. Amphetamines had no measurable effect on this behavior. Some of our preliminary studies indicated that this type of violent behavior may be decreased by diphenylhydantoin (Dilantin), but our data at this time are still very sketchy.

Comparative Effects of Methylphenidate (Ritalin) on Hyperkinesis

The methylphenidate data are of a preliminary nature, since we performed only three experiments on dog Jackson, using the dose levels of 2, 3, and 4 mg/kg. The dog's behavior was better on all three dose levels than in the no-drug state, and the dog discriminated well between the danger and safe signals. However, on the 2 and 3 mg/kg doses the dog had chewed and torn down all the equipment by the end of the signal period. The 4 mg/kg dose resulted in good behavioral control (i.e., no struggling, no biting or chewing, no crying), but the drug was not as "normalizing" as *d*-amphetamine: the dog had a "worried" look, showed some trembling, licking of lips, and some drug-induced repetitive movement of the head when inhibitory tones were presented. In contrast, when the dog was on *d*-amphetamine (1 mg/kg) he looked pleased, alert, "normal," and "showy."

The Effects of Amphetamines on Conditional Emotional Visceral Responses

As mentioned above, some of our dogs (referred to as antidiuretic), after exposure to a Pavlovian conditioning paradigm with electrocutaneous reinforcement, developed almost inextinguishable conditional emotional visceral re-

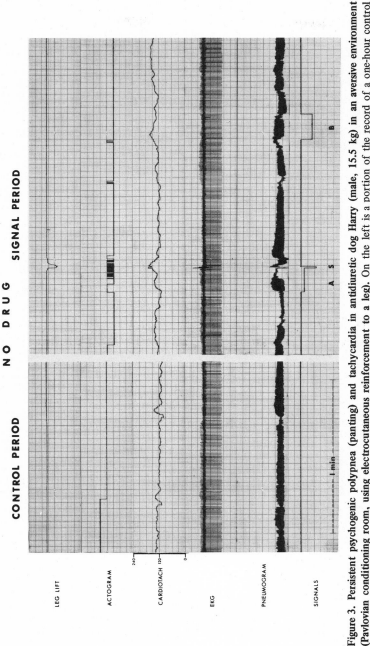

Figure 3. Persistent psychogenic polypnea (panting) and tachycardia in antidiuretic dog Harry (male, 15.5 kg) in an aversive environment (Pavlovian conditioning room, using electrocutaneous reinforcement to a leg). On the left is a portion of the record of a one-hour control period during which baseline data are collected. On the right is a portion of the recording during a 20-min signal period during which 10 excitatory (A) tones (400 Hz) and 10 inhibitory (B) tones (1000 Hz) are presented in a random sequence. Each A tone lasted for 10 sec, the last 0.5 sec of which is reinforced with an electrocutaneous stimulus (S) to a leg, using a constant current generator (10 mA). Each B tone was presented for 10 sec and was not reinforced with shock. Leg lift = motor defense response. Note the absence of distinct conditional cardiac or motor defense responses.

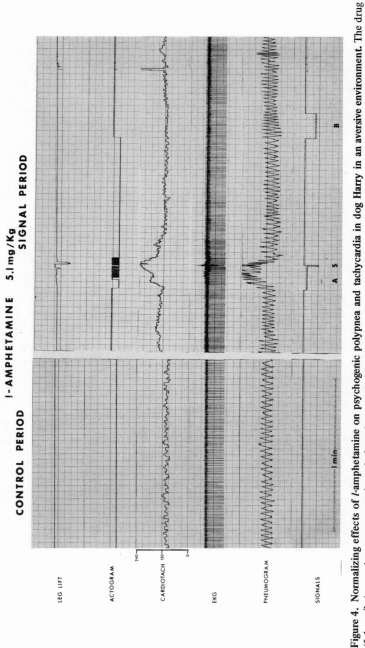

Figure 4. Normalizing effects of *l*-amphetamine on psychogenic polypnea and tachycardia in dog Harry in an aversive environment. The drug (5.1 mg/kg) was given per os one hour before the experiment. Designations are the same as in Figure 3. Note the appearance of distinct conditional cardiac, respiratory, and motor defense responses under the influence of amphetamine.

sponses to the entire conditioning room complex. Figure 3 is a photocopy of a portion of a record of responses of such an antidiuretic dog—Harry. We previously reported (Corson, 1971; Corson & Corson, 1971) that moderate dosages of antianxiety drugs (e.g., meprobamate, phenobarbital, diazepam) inhibited these psychogenic visceral reactions without affecting conditional Pavlovian motor defense or operant avoidance responses.

As can be seen in Figure 4, l-amphetamine (5.1 mg/kg) markedly reduced the tachycardia and polypnea during the control period. The heart rate was reduced from about 120 beats/min to about 90. The respiration was reduced from about 300/min to about 27. The polypnea was absent even during the signal period. Moreover, in a no-drug situation, the animal exhibited so much visceral turmoil that no distinct conditional motor defense, cardiac, or respiratory responses were apparent. Under the influence of amphetamine, clearly delineated conditional responses appeared.

Similar effects were observed with d-amphetamine, except that the dose required for visceral–autonomic normalization was 1.0–1.3 mg/kg, i.e., the dosage ratio of l:d was about 4.

The Problem of Tolerance to Amphetamine

No tolerance was observed in any of the dogs to repeated administration of either d- or l-amphetamine, in regard to the control of hyperkinesis or aggression. On the contrary, we often observed sensitization (reverse tolerance) to these drugs. For example, initially in dog Dunco the minimal dose required to control the hyperkinesis was 1.0 mg/kg. After several experiments, it was found that 0.2 mg/kg was sufficient to produce the same effects. Similar sensitization effects were observed by us in other hyperkinetic dogs. In contrast, normal dogs exhibited remarkable tolerance to repeated administration of amphetamines.

EPILOGUE AND OVERVIEW

The so-called "paradoxical" calming effect of amphetamines on hyperkinetic children has been described by Laufer and Denhoff (1957), Conners et al. (1967), and many others, and was recently extensively and elegantly reviewed by Wender (1971). Unfortunately, not all hyperkinetic children benefit from this therapy. As pointed out by Fish (1971), Twitchell (1971), and Wender (1971), this disorder is a sort of diagnostic wastebasket incorporating many dysfunctions, the nature of which is unknown. Numerous hypotheses have been proposed about the etiology of this group of disturbances and the mechanisms of beneficial action by amphetamines, but thus far no experimental verification has been forthcoming.

In order to elucidate the nature of this group of behavioral disturbances, experimenters have been looking for suitable animal models. Such a model of naturally occurring hyperkinetic dogs and cats is reported in this presentation.

Of the eight hyperkinetic dogs we investigated, six responded favorably to both *d*- and *l*-amphetamine. This happens to correspond roughly to the proportion of hyperkinetic children improved by amphetamine. This correspondence, however, is purely coincidental, since our sample of dogs studied is very small.

One of the significant factors determining the type of drug response obtained in dogs is probably genetic. The two dogs (one hyperkinetic and one with a learning disability) which did not respond to amphetamines are wirehair fox terriers. The hyperkinetic dogs responding favorably to amphetamines had a telomian, cocker spaniel, or English shepherd background. We propose to develop further investigations on the genetic aspects of this problem. It is important to emphasize that not all wirehair fox terriers are hyperkinetic. In fact, many of these dogs are easily trained and conditioned. It is, therefore, imperative that we breed the particular wirehair fox terriers which are hyperkinetic.

It is instructive to point out that in the control of hyperkinesis and of conditional emotional reactions (as well as in the production of stereotypy and anorexia) it was necessary to use 3–4 times as much *l*-amphetamine as of the *d*-isomer. In contrast, the two isomers were equipotent in the control of aggression in dog Jackson. If the intriguing interpretation proposed by Coyle and Snyder (1969), Snyder et al. (1970), and Taylor and Snyder (1970) is valid, then our findings would suggest a predominantly dopaminergic system involved in aggression and a predominantly noradrenergic system in hyperkinesis, stereotypy, and anorexia. This conclusion is predicated on the reports by Axelrod (1971) to the effect that amphetamine produces its effects by releasing catecholamines and blocking their reuptake.

However, recent reports by Svensson (1971), Harris and Baldessarini (1973), Ferris, Tang, and Maxwell (1972), and Thornburg and Moore (1973) have thrown doubt on the findings and interpretations of Coyle and Snyder. Harris and Baldessarini (1973) reported that *d*-amphetamine was four times more potent than the *l*-isomer in inhibiting the reuptake of dopamine in synaptosomes from striatal tissues of rats. Conversely, in cerebral cortex synaptosomal preparations there was less than a two-fold difference in the potencies of the two isomers. These findings are similar to those reported by Ferris et al. (1972) and by Thornburg and Moore (1973).

These findings would explain in part the reported four-fold differences in potency of *d*- and *l*-amphetamine in inducing stereotypy in rats (Taylor and Snyder, 1970; Scheel-Krüger, 1972), the *d*-isomer being the more potent. These data also fit in with the reports by Carlsson, Fuxe, Hamberger, and Lindquist (1966), Randrup and Munkvad (1967, 1970), Randrup and Scheel-Krüger (1966) and Scheel-Krüger (1972) to the effect that a dopaminergic system is involved in stereotyped behavior.

The reasons for these divergent findings are not clear. If one accepts the findings and interpretations of Svensson; Harris and Baldessarini; Ferris et al.; and Thornburg and Moore, then our own data on dogs would suggest that *aggressive behavior may involve chiefly a predominantly norepinephrine system, whereas hyperkinesis, stereotypy, and anorexia may involve primarily dopaminergic neurotransmitters.* This complex problem of differential psychobiologic effects of *d-* and *l-*amphetamine has recently been elegantly reviewed by Patil, Miller, and Trendelenburg (1974).

In this connection one should mention the interesting studies of Engel (1972) which suggest that dopamine is essential for elementary motor functions, whereas more complex integrated behavioral acts involve also norepinephrine neurotransmitters. Moreover, our experimental data throw no light on the possible interaction with cholinergic and serotonergic transmitters. Further studies on our experimental dogs are certainly indicated.

It is interesting that the learning which occurred under the influence of amphetamines in our hyperkinetic dogs was *not state-dependent*, i.e., all the conditional responses remained intact in no-drug situations. It is also important to note that the very first dose of amphetamine eliminated completely the hyperkinetic behavior and the dogs could be trained rapidly and were able to develop conditional discrimination after only 4–6 experimental sessions. This would support the suggestion by D. R. Meyer (personal communication; see also Braun, Meyer, & Meyer, 1966, and Jonason, Lauber, Robbins, Meyer, & Meyer, 1970) that amphetamines may facilitate primarily the process of information retrieval.

One of the most significant observations on the amphetamine effects on the violent behavior of the dog Jackson is related to the fact that after some six weeks of amphetamine and psychosocial therapy, the aggressive behavior largely disappeared after the drug was discontinued. It has now been about two years since the last dose of amphetamine was given to Jackson. He is still friendly and well behaved. Our tentative hypothesis is that amphetamine enabled the dog to develop adaptive social interactions which seemed to persist even after the drug was discontinued. This postulate would have to be tested by genetic breeding experiments, controlled postnatal experience, including exposure of littermates to different degrees of socialization, and the administration of drugs with and without opportunities for positive social interactions.

The fact that an animal with a genetic predisposition to violence can be (what appears to be permanently) transformed into an animal with positive social interactions has significant and hopeful implications for human violence, insofar as one is justified in extrapolating from animal studies to human social interactions. It certainly warrants the expansion of research on such animal models of nonadaptive violent behavior.

In this connection it may be instructive to note that a drug-initiated self-perpetuating *psychosocial virtuous cycle* observed in our experimental dogs would appear to be a mirror image of a *psychosocial vicious cycle* often initiated by minimal brain dysfunction symptoms in children, as described by Arnold (1973).

Another finding of interest is the dramatic inhibition of psychogenic autonomic-visceral responses in dog Harry by *d*- and *l*-amphetamine, the dosage ratio of *d*- to *l*- being 1:4. Barbeau (1970, 1971, 1972) made the intriguing suggestion that the striatum and a dopaminergic system may be involved in the modulation of feedback mechanisms regulating homeostatic functions of the autonomic ergotropic–trophotropic system.

Our work points out the importance of studying the effects of psychotropic drugs on animal models of psychopathology rather than merely on normal animals. Our naturally occurring hyperkinetic dogs and the dogs exhibiting psychovisceral pathology should prove useful in elucidating mechanisms of action of neurotropic drugs. For a discussion of the possible neurophysiologic mechanisms involved in the dogs with psychovisceral pathology see Corson et al. (1970) and Anokhin (1974).

The point we wish to convey is that the effects of amphetamines on certain kinds of hyperkinesis and aggression may *not really be paradoxical but may actually be normalizing* effects, in the same way as digitalis is beneficial in patients with congestive heart failure and detrimental in individuals with normal cardiac function. Supporting evidence for this concept is given by Knopp, Arnold, Smeltzer, and Andras (1972) and Knopp, Arnold, and Messiha (1973) in their studies on the interrelationship between the normalizing behavioral effects and the normalization of the extent of pupillary contraction in patients with Tourette's disease treated with haloperidol and in hyperkinetic children treated with amphetamine.

Similar conclusions were arrived at by Wikler, Dixon, and Parker (1970) on the basis of extensive EEG, neurological, and psychometric data and by Satterfield and Dawson (1971) and Satterfield (1973) on the basis of EEG and skin conductance studies in hyperkinetic children.

We postulate that the primary defect in uncontrollable hyperkinetic amphetamine responders (animal or children) may be related to a deficiency in the functioning of inhibitory systems. In terms of Pavlovian typology, such organisms would be classified as "choleric" or "melancholic." The hyperactivity is only one particular resultant of the hypoinhibitory defect.

As pointed out by Wender (1971), Anderson (1946) and Bender (1949), some children with Minimal Brain Dysfunction (MBD) are actually motorically *hypoactive* and listless but they present many of the other symptoms of the MBD syndrome, and many of them become normalized by amphetamine. Many MBD children appear to be hyperkinetic primarily because of their *shifting non-goal-directed activities and short attention span*. As stated by Wender: "These children are no more active than other children on the playground but cannot curtail their activity in the classroom."

Preliminary observations by Andreasen, Peters, and Knott (1975) on contingent negative variation (CNV) in seven hyperkinetic children (8-12 years old) and seven matched controls support the notion that the primary defect in hyperkinetic children is related primarily to a decreased attention span. Walter, Cooper, Aldridge, McCallum, and Winter (1964) and Walter, Cooper, Crow, Mc-Callum, Warren, Aldridge, Storm van Leeuwen, and Kamp (1967) suggested that the CNV is related to conation or an "expectancy" or "readiness" preceding the performance of a task. Generally, the CNV tends to be poorly developed in children below the age of 8. Andreasen et al. report that CNV waves "were not present in most of the hyperkinetic children while off medication, but when on cerebral stimulants their CNVs were similar to those of the controls."

All these observations are in agreement with our studies on our hyperkinetic dogs. In their own kennels the hyperkinetic dogs cannot be easily distinguished from normal easily trainable and conditionable dogs. The hyperkinetic defect becomes apparent particularly when these dogs are placed in a "Pavlovian conditioning school." They will not tolerate or become accommodated to the mild restraints of the Pavlovian stand and they appear to be incapable of filtering information or of responding to positive or negative reinforcement.

For these reasons, it would be more appropriate to drop the expression *hyperkinetic syndrome* and to adopt the term *hypoinhibitory syndrome*. This suggestion may have more than a mere epistemological significance. It may help to devise appropriate experimental approaches to the solution of this problem. It may also help to improve diagnostic approaches and may furnish teachers and parents more useful criteria for distinguishing between true behavioral pathology and children who are simply normally boisterous or who may become disruptive and hyperactive in a classroom because the level of instruction may be too high or too low for their particular intellectual and achievement level.

The focussing of attention on the hypoinhibitory defect, rather than on the hyperactivity, may also help to dispel irrational fears and opposition to rational selective drug therapy of the hypoinhibitory syndrome on the part of parents (especially in economically deprived social strata) who may look upon drug therapy as an attempt to "dope" their children and to prevent the expression of opposition to socioeconomic injustice. Such absolute dogmatic opposition to any drug therapy has been voiced even by professionally competent psychologists, psychiatrists, and educators. Such attitudes stem from failure to distinguish between normal adaptive boisterous active behavior and psychopathologic disturbances occasioned by biochemical neurotransmitter defects requiring specific therapy which may involve the use of drugs.

We postulate that the hypoinhibitory syndrome may be due to a deficiency in catecholamine transmitters (or a decrease in sensitivity to those transmitters) in inhibitory brain systems. Amphetamines, in appropriate dosages, may correct this defect by supplying this catecholamine transmitter deficiency.

Support for this hypothesis is furnished by our observations that our hyperkinetic dogs became normalized only by low dosages of amphetamines (0.2–1.0 mg/kg of the *d*-isomer). Presumably these are the dosages just sufficient for the liberation of deficient neurotransmitters in the inhibitory centers. Higher dosages failed to relieve the hyperkinesis and induced stereotypic movements, presumably by "flooding" all synaptic areas with released catecholamines.

In children in most cases, this neurotransmitter deficiency in inhibitory systems may be a maturational problem, since most children (though not all) seem to outgrow the syndrome during late adolescence. In our dogs this hypoinhibitory transmitter defect appears to be fixed and unchanged with maturation and to be associated with genetic factors. These naturally occurring animal models are therefore excellent material for elucidating the neurochemical and neurophysiologic mechanisms of this disorder and to develop better diagnostic and rational therapeutic and prophylactic methods.

In conclusion it should be mentioned that it is not possible to convey verbally the dramatic modification of hypoinhibitory and violent behavior by the amphetamines. We therefore prepared a documentary film (Corson, Corson, Kirilcuk, Arnold, Knopp, and Kirilcuk, 1974).

SUMMARY

In spite of the plethora of literature on minimal brain dysfunction or the hyperkinetic syndrome, very little is known about the basic causes of this childhood disability. Similarly, there is much confusion about diagnostic and prognostic criteria or rational therapeutic methods. Three major recent documents (Freedman, 1971; Wender, 1971; and *Annals N.Y. Academy of Sciences*, 1973, 205, 1–396, containing papers presented at a conference on Minimal Brain Dysfunction in 1972), representing the accumulated experience and considered judgement of expert clinicians and researchers, all testify to the current lack of a scientific basis for the development of rational diagnostic and therapeutic criteria.

The current "state of the art" was succinctly summarized by Masland in the Epilogue to the above-mentioned N.Y. Academy of Sciences monograph (1973):

> This monograph again highlights the diversity of problems that are subsumed by the terms "minimal brain dysfunction" (MBD) or "specific learning disability" (SLD). These are not diagnostic terms. We are not speaking of a disease. We are merely recognizing that there is a group of individuals who have certain characteristics in common, as a result of which they share a need for certain special management or management facilities. It is a great error to feel that when you have labeled somebody MBD or SLD, you have made a diagnosis. All you have really said is that there is something different about this individual's intellectual characteristics, and he is likely to require special attention.

The current state of knowledge of the etiology and drug therapy of MBD is comparable to the early indiscriminate uses of digitalis in the treatment of edema following the publication of the monograph by William Withering in 1785 before it was clearly established that the main value of digitalis is in the therapy of cardiac edema.

Similar confusion exists in relation to the psychobiology, therapy, and prevention of uncontrollable violent behavior in some hyperkinetic children and in adults.

The major obstacle to the solution of these problems has been the lack of suitable animal models of these behavioral disturbances. For the first time such naturally occurring animal models (dogs and cats) have been discovered and studied in our laboratory. The behavior of these animal models exhibits characteristics comparable to those described for MBD children:

1. These animals exhibit uncontrollable hyperkinetic behavior when placed in a Pavlovian conditioning situation, a situation comparable to the discipline-requiring school environment.

2. These animals fail to respond to reward or to punishment; they are negativistic, disobedient, and obstinate.

3. Some of these animal models are also impulsive, have a low frustration tolerance, have a deficiency in delaying gratification, and may exhibit incorrigible violent nonadaptive behavior.

4. Some of these animal models (but not all) respond favorably to amphetamines in dosages comparable to those found useful in some MBD children.

5. Under amphetamine therapy, these dogs become easily trainable, and can acquire rapidly discriminated classical and operant conditioned responses.

6. The learning acquired under amphetamine therapy remained intact following drug withdrawal.

7. Some of the hyperkinetic dogs exhibiting incorrigible violent behavior also responded favorably to amphetamines, the normalization persisting after drug withdrawal.

8. The MBD syndrome in most children (but not in all—see Wender, 1971) appears to be a maturational problem, with a suggestion of genetic predisposition. In our animal models this disturbance appears to be fixed and also has a

strong indication of genetic predisposition. Therefore genetic breeding of these animal models combined with controlled postnatal experiences may give us more definite indications as to the interaction between genetics and early experiential factors in determining nonadaptive hyperkinetic and violent behavior.

9. Our experiments suggest that perhaps the primary defect in hyperkinetic and violent behavior in our animals is related to neurotransmitter disturbances in inhibitory systems in the brain. This suggestion can be subjected to experimental verification.

Thus, these animal models represent suitable subjects for elucidating the physiologic and neurohumoral mechanisms underlying the several diagnostic categories of hyperkinetic and violent behavior and for developing rational diagnostic, prognostic, and therapeutic criteria which might be applicable to MBD children and to humans (children and adults) exhibiting various types of nonadaptive violent behavior. The fact that we have animal MBD models, some of which do not respond to amphetamines, makes these models even more relevant to the MBD problem in children.

Our findings indicate that a combination of psychosocial and drug therapy can transform a genetically predisposed violent organism into a peaceful, cooperating, lovable animal. Insofar as one is justified in extrapolating from animal experiments to human affairs, this discovery has significant and hopeful implications for the possibilities of modifying violent and sociopathic behavior in children and in adults. Expanded experimental studies on our animal models of violence would certainly represent a worthwhile investment in the development of peaceful and cooperative social interactions.

ACKNOWLEDGMENT

This research was supported in part by USPHS grants MH 12089, MH 18098, Biomed. Sci. Support Grant RR-07074, and the Grant Foundation, Inc.

REFERENCES

Anderson, C. Early brain injury and behavior. *J. Amer. Med. Wom. Ass.*, 1956, *11*, 113–119.
Andreasen, N. J. C., Peters, J. F., and Knott, J. R. CNVs in hyperactive children: effects of chemotherapy. In W. C. McCallum and J. R. Knott (Eds.), Proceedings of Third International Congress on Event Related Slow Potentials of the Brain, Bristol, 1975.
Anokhin, P. K. Biology and Neurophysiology of the Conditioned Reflex and Its Role in Adaptive Behavior. S. A. Corson, Scientific and Translation Editor. International Series of Monographs on Cerebrovisceral and Behavioral Physiology and Conditioned Reflexes (Vol. 3), Oxford and New York: Pergamon Press, 1974.

Arnold, L. E. Is this label necessary? *Journal of School Health*, 1973, *23(8)*, 510–514.

Arnold, L. E., Kirilcuk, V., Corson, S. A., & Corson, E. O'Leary. Levoamphetamine and dextroamphetamine: differential effect on aggression and hyperkinesis in children and dogs. *Amer. J. Psychiat.*, 1973, *130(2)*, 165–170.

Arnold, L. E., Wender, P. H., McCloskey, K., & Snyder, S. Comparative efficacy in the hyperkinetic syndrome; assessment by target symptoms. *Arch. Gen. Psychiat.*, 1972, *27*, 35–41.

Axelrod, J. Noradrenaline: fate and control of its biosynthesis. *Science*, 1971, *173*, 598–606.

Barbeau, A. (1969). L-Dopa therapy in Parkinson's Disease: a critical review of nine years' experience. *Can. Med. Ass. J.*, 1969, *101*, 971–800.

Barbeau, A. Dopamine and disease. *Can. Med. Ass. J.*, 1970, *103*, 824–832.

Barbeau, A. Functions of the striatum. In De Ajuriaguerra (Ed.), The Fourth Bel-Air Symposium, Geneva: Masson, 1971.

Barbeau, A. Role of Dopamine in the nervous system. In: Proceedings of the Third International Congress on Neuro-Genetics and Neuro-Ophthalmology, Brussels, 1970. Basel: S. Karger, 1972, pp. 114–136.

Bender, L. Psychological problems of children with organic brain disease. *Amer. J. Orthopsychiat.*, 1949, *19*, 404–415.

Bradley, C. The behavior of children receiving benzedrine. *Amer. J. Psychiat.*, 1937, *94*, 577–585.

Bradley, C. Benzedrine and Dexedrine in the treatment of children's behavior disorders. *Pediatrics*, 1950, *5*, 24–36.

Bradley, C. Management of behavior problems in children. *Med. Clin. N. Amer.*, 1953, *37*, 565–577.

Braun, J. J., Meyer, P. M., & Meyer, D. R. Sparing of a brightness habit in rats following visual decortication. *J. Comp. Physiol. Psychol.*, 1966, *61*, 79–82.

Burks, H. F. Effects of amphetamine therapy on hyperkinetic children. *Arch. Gen. Psychiat.*, 1964, *11*, 604–609.

Cannon, W. B. *The Wisdom of the Body*. New York: W. W. Norton and Company, 1932.

Carlsson, A., Fuxe, K., Hamberger, B., & Lindqvist, M. Biochemical and histochemical studies of the effects of imipramine-like drugs and (+)-amphetamine on central and peripheral catecholamine neurons. *Acta physiol. Scand.*, 1966, *67*, 481–497.

Clements, S. D., & Peters, J. E. Minimal brain dysfunctions in the school-age child. Diagnosis and treatment. *Arch. Gen. Psychiat.*, 1962, *6*, 185–197.

Conners, C. K., & Eisenberg, L. The effects of methylphenidate on symptomatology and learning in disturbed children. *Amer. J. Psychiat.*, 1963, *120*, 458–464.

Conners, C. K., Eisenberg, L., & Barcai, A. Effect of dextroamphetamine on children. *Arch. Gen. Psychiat.*, 1967, *17*, 478–485.

Conners, C. K., Eisenberg, L., & Sharpe, L. Effects of methylphenidate (Ritalin) on paired-associate learning and Porteus maze performance in emotionally disturbed children. *J. Consult. Psychol.*, 1964, *28(1)*, 14–22.

Conrad, W. G., & Insel, J. Anticipating the response to amphetamine therapy in the treatment of hyperkinetic children. *Pediatrics*, 1967, *40(1)*, 96–99.

Corson, S. A., Corson, E. O'Leary, & England, S. J. M. The influence of meprobamate on conditioned and unconditioned visceral and motor defense responses. In Psychopharmacological Methods, London: Pergamon Press, 1963, pp. 244–255.

Corson, S. A. Conditioning of water and electrolyte excretion. In Endocrines and the Central Nervous System. *Res. Publ., Ass. Res. Nerv. Ment. Dis.*, Baltimore: Williams and Wilkins, 1966a, *43*, 140–199.

Corson, S. A. Neuroendocrine and behavioral response patterns to psychologic stress and the problem of the target tissue in cerebrovisceral pathology. Proceedings of the Conference on Psychophysiological Aspects of Cancer, *Ann. N.Y. Acad. Sci. 125*, 1966b (Article 3), 890–918.

Corson, S. A., & Corson, E. O'Leary. Pavlovian conditioning as a method for studying the mechanisms of action of minor tranquilizers. In H. Brill, (Ed.), Neuropsychopharmacology. *Exerpta Med. Found., Int. Congr. Ser.*, 1967, no. 129, 857–881.

Corson, S. A. Physiologic responses to avoidable and unavoidable psychologic stress in relation to genetic differences. *Ann. N.Y. Acad. Sci.*, 1969, *164*, 526.

Corson, S. A. Pavlovian and operant conditioning techniques in the study of psychosocial and biological relationships. In L. Levi (Ed.), Society, Stress and Disease (Vol. 1), The Psychosocial Environment and Psychosomatic Diseases, London: The Oxford University Press, 1971, pp. 7–21.

Corson, S. A., & Corson, E. O'L. (1971): Psychosocial influences on renal function—implications for human pathophysiology. In L. Levi (Ed.), Society, Stress and Disease (Vol. 1), The Psychosocial Environment and Psychosomatic Diseases, London: The Oxford University Press, 1971, pp. 338–351.

Corson, S. A., Corson, E. O'L., & Kirilcuk, V. Individual differences in respiratory responses of dogs to psychologic stress and Anokhin's formulation of the functional system as a unit of biological adaptation. *Int. J. Psychobiol.*, 1970, *1*, 1–16.

Corson, S. A., Corson, E. O'L., Kirilcuk, V., & Arnold, L. E. Tranquilizing effects of d-amphetamine on hyperkinetic untrainable dogs. *Fed. Proc.*, 1971, *30*, 206.

Corson, S. A., Corson, E. O'L., Kirilcuk, V., Knopp, W., & Arnold, L. E. Differential interaction of amphetamines and psychosocial factors in the modification of violent and hyperkinetic behavior and learning disability. *Fed. Proc.*, 1972a, *31*, 820.

Corson, S. A., Corson, E. O'L., Kirilcuk, V., Arnold, L. E., & Knopp, W. Bridging the gap between psychobiology and medical practice—development of an animal model for the study of the hyperkinetic syndrome. In H. Goldman (Ed.), Research in Comprehensive Psychiatry: A Festschrift for Ralph Patterson, Columbus: Ohio State Univ. Press, 1972b.

Corson, S. A., Corson, E. O'Leary, Kirilcuk, V., Kirilcuk, J., Knopp, W., and Arnold, L. E. Differential effects of amphetamines on clinically relevant dog models of hyperkinesis and stereotypy: Relevance to Huntington's chorea. In A. Barbeau, T. N. Chase, & G. W. Paulson (Eds.), Advances in Neurology (Vol. 1), Huntington's Chorea, 1872–1972, New York: Raven Press, 1973, pp. 681–697.

Corson, S. A., Corson, E. O'L., Kirilcuk, V., Arnold, L. E., Knopp, W., & Kirilcuk, J. Experimental control of hyperkinetic and violent behavior in dogs. (16 mm B/W sound film, 38 minutes. Distributed by the Psychological Cinema Register, The Pennsylvania State University, Audio–Visual Services, 6 Willard Building, University Park, Pa. 16802) 1974.

Coyle, J. T., & Snyder, S. H. Catecholamine uptake by synaptosomes in homogenates of rat brain: Stereospecificity in different areas. *J. Pharmac. Exp. Ther.*, 1969, *170*, 221–231.

Creager, R. O., & Van Riper, C. The effect of methylphenidate on the verbal productivity of children with cerebral dysfunction. *J. Speech Hearing Res.*, 1967, *10(3)*, 623–628.

de la Cruz, F. F., Fox, B. H., & Roberts, R. H. (Eds.). Minimal brain dysfunction. *Annals of the New York Academy of Sciences*, 1973, Vol. 205, 396 pp.

Eisenberg, L. Psychiatric implications of brain damage in children. *Psychiat. Quart.*, 1957, *31(1)*, 72–92.

Eisenberg, L. Possibilities for a preventive psychiatry. *Pediatrics*, 1962, *30(5)*, 815–828.

Eisenberg, L. The management of the hyperkinetic child. *Develop. Med. Child Neurol.*, 1966, *8*, 593–598.

Eisenberg, L., Conners, C. K., & Sharpe, L. A controlled study of the differential application of outpatient psychiatric treatment for children. *Jap. J. Child Psychiat.*, 1965, *6(3)*, 125–132.

Eisenberg, L., Lachman, R., Molling, P. A., Lockner, A., Mizelle, J. D., & Conners, C. K. A psychopharmacologic experiment in a training school for delinquent boys: Methods, problems, findings. *Amer. J. Orthopsychiat.*, 1963, *33*, 431–447.

Ellinwood, E. H., Jr. Amphetamine psychosis: a multi-dimensional process. *Seminars Psychiat.*, 1969, *1*, 208–226.

Engel, J. Neurochemistry and Behavior. A Correlative Study with Special Reference to Central Catecholamines. Göteborg: University of Göteborg, 1972.

Epstein, L. C., Lasagna, L., Conners, C. K., & Rodriguez, A. Correlation of dextroamphetamine excretion and drug response in hyperkinetic children. *J. Nerv. Ment. Dis.*, 1968, *146(2)*, 136–146.

Ferris, R. M., Tang, F., & Maxwell, R. (1972). A comparison of the capacities of isomers of amphetamine, deoxypipradrol and methylphenidate to inhibit the uptake of tritiated catecholamines into rat cerebral cortex slices, synaptosomal preparations of rat cerebral cortex, hypothalamus and striatum and into adrenergic nerves of rabbit aorta. *J. Pharmac. Exp. Ther.*, 1972, *181*, 407–416.

Fish, B. The "One Child, One Drug" myth of stimulants in hyperkinesis. *Arch. Gen. Psychiat.*, 1971, *25*, 193–203.

Freedman, Alfred M. & Kaplan, Harold I. Comparative Textbook of Psychiatry, Baltimore: Williams and Wilkins Co., 1967, p. 1666.

Freedman, Daniel X. Report of the conference on the use of stimulant drugs in the treatment of behaviorally disturbed young school children. *Psychopharmacol. Bull.*, 1971, *7(3)*, 23–29.

Gerard, R. W. Concluding remarks: Meprobamate and other agents used in mental disturbances. *Ann. N.Y. Acad. Sci.*, 1957, *67(10)*, 885.

Harris, J. E., & Baldessarini, R. J. Uptake of [^{3}H]-catecholamines by homogenates of rat corpus striatum and cerebral cortex: effects of amphetamine analogues. *Neuropharmacology*, 1973, *12*, 669–679.

Huessy, H. R. Study of the prevalence and therapy of the choreatiform syndrome or hyperkinesis in rural Vermont. *Acta Paedopsychiat.*, 1967, *34*, 130–135.

Jasper, H. H., Solomon, P., & Bradley, C. Electroencephalographic analyses of behavior problem children. *Amer. J. Psychiat.*, 1938, *95*, 641–658.

Jenkins, R. Diagnostic classification in child psychiatry. *Amer. J. Psychiat.*, 1970, *127(5)*, 680–681.

Jonason, K. R., Lauber, S. M., Robbins, M. J., Meyer, P. M., & Meyer, D. R. Effects of amphetamine upon relearning pattern and black-white discriminations following neocortical lesions in rats. *J. Comp. Physiol. Psychol.*, 1970, *73*, 47–55.

Knights, R. M., & Hinton, G. G. The effects of methylphenidate (Ritalin) on the motor skills and behavior of children with learning problems. *J. Nerv. Ment. Dis.*, 1969, *148(6)*, 643–653.

Knopp, W., Arnold, L. E., Smeltzer, D. J., & Andras, R. L. Pupillary light reaction as a predictor of amphetamine response in hyperkinetic children. *Psychopharmacologia*, 1972, *26*, 53.

Knopp, W., Arnold, L. E., & Messiha, F. S. Tourette's disease: Implications for research in Huntington's disease. In A. Barbeau, T. N. Chase, & G. W. Paulson (Eds.), Advances in Neurology (Vol. 1), Huntington's Chorea, 1872–1972, New York: Raven Press, 1973, pp. 135–145.

Kornetsky, C. Psychoactive drugs in the immature organism. *Psychopharmacologia (Berlin)*, 1970, *17(2)*, 105–136.

Ladisich, W., Volbehr, H., & Matussek, N. Paradoxical amphetamine effect in hyperactive rats in relation to norepinephrine metabolism. *Neuropharmacology*, 1970a, *9(4)*, 303–310.

Ladisich, W., Volbehr, H., & Matussek, N. Paradoxical effect of amphetamine on hyperactive states in correlation with catecholamine-metabolism in brain. In E. Costa and S. Garattini (Eds.), Amphetamines and Related Compounds, New York: Raven Press, 1970b, pp. 487–492.

Laufer, M. W. Cerebral dysfunction and behavior disorders in adolescents. *Amer. J. Orthopsychiat.*, 1962, *32*, 501–506.

Laufer, M. W., & Denhoff, E. Hyperkinetic behavior syndrome in children. *J. Pediat.*, 1957, *50*, 463–473.

Laufer, M. W., Denhoff, E., & Solomons, G. Hyperkinetic impulse disorder in children's behavior problems. *Psychosomat. Med.*, 1957, *19(1)*, 38–49.

Masland, R. Testimony before a Subcommittee of the Committee on Appropriations. House of Representatives, 89th Congress, First Session, part 3, Wash., D.C.: U.S. Govt. Printing Office, 1965.

Meyer, D. R. Personal communication.

Millichap, J. G. Drugs in the management of hyperkinetic and perceptually handicapped children. *J. Amer. Med. Ass.*, 1968, *206(7)*, 1527–1530.

Millichap, J. G., Aymat, F., Sturgis, L. H., Larsen, K. W., & Egan, R. A. Hyperkinetic behavior and learning disorders. III. Battery of neuropsychological tests in controlled trial of methylphenidate. *Amer. J. Dis. Child*, 1968, *116*, 235–244.

Millichap, J. G., & Fowler, G. W. Treatment of "minimal brain dysfunction" syndrome. *Pediat. Clin. N. Amer.*, 1967, *14(4)*, 767–777.

Molitch, M., & Eccles, A. K. The effect of benzedrine sulfate on the intelligence scores of children. *Amer. J. Psychiat.*, 1967, *94*, 587–590.

Nichamin, S. J., & Barahal, G. D. Faulty neurologic integration with perceptual disorders in children. *Mich. Med.*, 1968, *67(17)*, 1071–1075.

Patil, P. N., Miller, D. D., & Trendelenburg, U. Molecular geometry and adrenergic drug activity. *Pharmacol. Rev.* 1974, *26(4)*, 323–392.

Patil, P. N., LaPidus, J. B., & Tye, A. Steric aspects of adrenergic drugs. I. Comparative effects of DL isomers and desoxy derivatives. *J. Pharmacol. Exp. Ther.*, 1967, 1–12.

Randrup, A., & Munkvad, I. Stereotyped activities produced by amphetamine in several animal species and man. *Psychopharmacologia (Berl.)*, 1967, *11*, 300–310.

Randrup, A., & Munkvad, I. Biochemical, anatomical and psychological investigations of stereotyped behaviour induced by amphetamines. In E. Costa, & S. Garattini, (Eds.), Amphetamines and Related Compounds, New York: Raven Press, 1970, pp. 695–713.

Randrup, A., & Scheel-Krüger, J. Diethyldithiocarbamate and amphetamine stereotype behaviour. *J. Pharm. Pharmac.*, 1966, *18*, 752.

Satterfield, J. H. EEG issues in children with minimal brain dysfunction. *Seminars in Psychiatry* 1973, *5(1)*, 35–46. Also reprinted in S. Walzer & P. H. Wolff (Eds.), Minimal Cerebral Dysfunction in Children, New York: Grune and Stratton, 1973.

Satterfield, James H., & Dawson, M. E. Electrodermal correlates of hyperactivity in children. *Psychophysiology*, 1971, *8*, 191–197.

Scheel-Krüger, J. Behavioural and biochemical comparison of amphetamine derivatives, cocaine, benztropine and tricyclic anti-depressant drugs. *Eur. J. Pharmac.*, 1972, *18*, 63–73.

Scott, J. P., & Fuller, J. L. Genetics and the Social Behavior of the Dog. Chicago: University of Chicago Press, 1965.

Snyder, S. H., Taylor, K. M., Coyle, J. T., & Meyerhoff, J. L. The role of brain dopamine in behavioral regulation and the actions of psychotropic drugs. *Amer. J. Psychiat.*, 1970, *127*(2), 199–207.

Stockard, C. R., Anderson, O. D., & James, W. T. The genetic and endocrinic basis for differences in form and behavior. In American Anatomical Memoirs, No. 19, Philadelphia: Wistar Institute of Anatomy and Biology, 1941.

Svensson, T. H. Functional and biochemical effects of *d*- and *l*-amphetamine on central catecholamine neurons. *Naunyn-Schmiedebergs Arch. Pharmak.*, 1971, *271*, 170–180.

Taylor, K. M., & Snyder, S. H. Amphetamine: differentiation by *d* and *l* isomers of behavior involving brain norepinephrine or dopamine. *Science*, 1970, *168*, 1487–1489.

Thornburg, J. E. & Moore, K. E. Dopamine and norepinephrine uptake by rat brain synaptosomes: relative inhibitory potencies of l- and d-amphetamine and amantadine. *Res. Comm. Chem. Pathol. Pharmacol.*, 1973, *5*(*1*), 81–89.

Twitchell, T. E. A behavioral syndrome. *Science*, 1971, *174*, 135–136.

Voeller, K. K. S., & Kempf, John. Personal communication.

Wallach, M. B., & Gershon, S. (1972). The induction and antagonism of central nervous system stimulation-induced stereotyped behavior in the cat. *Eur. J. Pharmac.*, 1972, *18*, 22–26.

Walter, W. Grey, Cooper, R., Aldridge, V. J., McCallum, W. C., & Winter, A. L. Contingent negative variation: an electric sign of sensorimotor association and expectancy in the human brain. *Nature*, 1964, *203*(*4943*), 380–384.

Walter, W. Grey, Cooper, R., Crow, H. J., McCallum, W. C., Warren, W. J., Aldridge, V. J., Storm van Leeuwen, W., & Kamp, A. Contingent negative variation and evoked responses recorded by radio-telemetry in free-ranging subjects. *Electroencephalog. Clin. Neurophysiol.*, 1967, *23*(*3*), 197–206.

Wender, P. H. *Minimal Brain Dysfunction in Children*. New York: Wiley-Interscience, 1971.

Wikler, A., Dixon, J. F., & Parker, J. B., Jr. (1970). Brain function in problem children and controls: psychometric, neurological, and electroencephalographic comparisons. *Amer. J. Psychiat.*, 1970, *127*(*5*), 634–645.

Coping Behavior and Neurochemical Changes
An Alternative Explanation for the Original "Learned Helplessness" Experiments

J. M. WEISS, H. I. GLAZER, AND
L. A. POHORECKY

INTRODUCTION

For several years, we have been investigating the question of how psychological variables in stress situations affect physiological responses. This is shown diagrammatically in Figure 1 as the influence of (A) on (B). It is of particular relevance for mental/behavioral disorders if such investigations can be carried a step further, which is to determine whether the differential physiological changes (B) that arise because of psychological variables in stress situations (A) can in turn mediate subsequent behavioral changes (C). In this event, we shall be able to observe how significant psychological factors lead to behavioral change via physiological mediation that is known to us.

The possibility of constructing such a model presents itself from studies that have examined the effects of controlling a stressor, or in other words, the effects of coping. The influence of this particular physiological variable has been studied intensively in our laboratory. These studies of coping have consisted of compar-

J. M. WEISS, H. I. GLAZER, and L. A. POHORECKY ● The Rockefeller University, New York, New York.

ing the effects that occurred when an animal could avoid or escape electric shocks with those effects that occurred when an animal received the same shocks but could not control them. Such experiments have revealed that the psychological dimension of stressor control is a major determinant of a variety of physiological stress responses. When animals could not control shock, they showed, in relation to those that received the same shocks while exercising control, more severe gastric lesions, greater loss of body weight, higher plasma steroid levels, and more fearfulness (Weiss, 1968; 1971a–c). These effects are shown in the lower portion of Figure 1, with the relationship between differences in coping ability and the physiological consequences thereof indicated by the first solid arrow on the left.

At the same time as we were studying the physiological effects of coping, other laboratories were studying behavioral changes produced by this same psychological factor. Overmier and Seligman (1967) and Seligman and Maier (1967) reported that dogs failed to learn and performed a shuttle avoidance response following exposure to electric shock over which they had no control, whereas dogs exposed to similar shocks while performing an avoidance–escape response showed no impairment in subsequent avoidance performance. Richardson, Scudder, and Karczmar (1970) reported that mice which had received inescapable shock showed less general activity and less exploratory behavior in an open field than did mice which received similar shock in the course of an avoidance task. Thus, the ability or inability to perform a coping response has been reported to affect behavioral as well as physiological responses, with inescapable shock apparently leading to poor avoidance–escape learning and reduced motor activity in comparison with escapable and/or avoidable shock. The major purpose of the present paper is to consider whether the behavioral effects that result

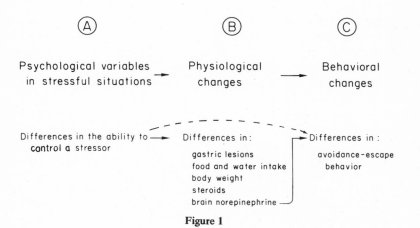

Figure 1

from differences in the ability to control a stressor are related to the physiological changes that also result from the same stressful conditions.

"Learned Helplessness" Hypothesis

According to the investigators (Overmier & Seligman, 1967; Seligman & Maier, 1967) who initially reported deficits in avoidance behavior following inescapable (but not escapable) shock, there is no necessary connection between stress-induced physiological changes and the behavioral (avoidance–escape) deficit they observed. These investigators explained their behavioral results as follows: animals exposed to shock which they could not control learned that their behavior did not alter their environment; therefore, having learned that "nothing I do matters," they subsequently failed to acquire an active avoidance response when training was later attempted. On the other hand, animals exposed to shock which they could control did not learn to be "helpless" and therefore could acquire the new avoidance response. Seligman and co-workers contend that the phenomenon of "giving up," which they call "learned helplessness," was not only present in their dogs but also (a) has been seen in a variety of other species in a number of situations although the experimenters have not recognized the phenomenon as such (Seligman, Maier, & Solomon, 1971); (b) has been demonstrated in rats (Maier, Albin, & Testa, 1973; Seligman and Beagley unpublished manuscript); and (c) underlies certain forms of psychopathology in humans (Seligman, 1973). In Figure 1, the "learned helplessness" relationship, which does not rely on any of the stress-induced physiological changes such as those mentioned under (B), is indicated by the dotted arrow.

The behavior of the dogs after inescapable shock was important in suggesting the "learned helplessness" explanation for the avoidance–escape deficit, and therefore this behavioral pattern should be described. In the shuttle box, the dogs were required to cross from one side of the box to the other (by jumping a hurdle) in order to terminate the warning signal and/or the shock which followed the signal. Although the dogs that had previously received inescapable shock appeared fully capable of executing the correct response, instead they moved around the box during the initial shock(s) without jumping the hurdle until they eventually stopped moving about and simply howled for the remainder of the 60-sec trial. On succeeding trials, their activity progressively decreased until they were generally immobile throughout most of the time that the shock was on. Also, if the animal did make an escape response, it failed to consistently repeat this success on subsequent trials. On the other hand, control animals—both those that never received shock previously or those that had been able to escape shock in the earlier experience—likewise moved around the box during the initial shock(s) but eventually jumped the hurdle and quickly learned to do this whenever the signal and/or the shock occurred.

"Motor-Activation Deficit" Hypothesis

In contrast to the learned-helplessness hypothesis, it has been proposed that the behavioral deficit observed by Overmier and Seligman is indeed directly related to stress-induced physiological changes. Weiss, Krieckhaus, and Conte (1968) and Miller and Weiss (1969) suggested that the avoidance deficit was produced by some form of physiological debilitation arising from the inescapable shock condition. Important evidence for this explanation came from one of the experiments conducted by Overmier and Seligman (1967). Overmier and Seligman reported that poor avoidance performance in the dogs was evident 24 hours after a session of inescapable shock but was totally absent if the dogs were first tested 48 hours after shock. This highly important result, showing that the avoidance–escape deficit disappeared if more than one to two days intervened between the inescapable shock session and the avoidance–escape test, was replicated by Overmier (1968). According to Weiss et al. (1968) and Miller and Weiss (1969), this type of time course (rapid dissipation of the effect) is not at all characteristic of learned responses, and thus the experiment by Overmier and Seligman does not support the contention that learned helplessness produced the deficit. Rather, proposed Weiss, Miller, and co-workers, the time course suggests that inescapable shock induced some form of physiological imbalance which corrects itself with the passage of time.

Shortly thereafter, Weiss, Stone, and Harrell (1970) outlined an alternative explanation, which hereafter is referred to as the "motor-activation deficit" hypothesis. In the experiment carried out by Weiss et al. (1970), it was observed that animals receiving inescapable shock subsequently had lower levels of norepinephrine in the brain than did animals receiving exactly the same shocks while performing an avoidance–escape response. Since norepinephrine appeared to be important in mediating active motor behavior—both motor activity (Herman, 1970; Segal & Mandell, 1970) and active avoidance responding (Moore, 1966; Rech, Bovys, & Moore, 1966; Seiden & Peterson, 1968)—it was suggested that changes in central noradrenergic activity produced by differences in coping behavior could account for the avoidance differences observed in the dog experiments. This hypothesis can be basically stated as follows. When animals were exposed to severe inescapable shock, a deficiency in central noradrenergic activity occurred, deriving, in part at least, from the depletion in the level of norepinephrine seen in this condition. As a consequence of this noradrenergic deficiency, the animals could mediate only a limited amount of motor activity, an amount insufficient for learning and performance of the correct response in the shuttle-avoidance task on which they were tested. On the other hand, animals exposed to the same shocks but able to control them did not develop this noradrenergic deficiency and therefore could mediate sufficient motor behavior to perform adequately. This hypothesis explains why all animals initially attempted to escape from the shock and why those who received inescapable shock soon ceased these attempts. In Figure 1, this hypothesis is represented as the progression indicated by the two solid arrows.

RECENT NEUROCHEMICAL FINDINGS IN REGARD TO COPING BEHAVIOR

Since the original study by Weiss et al. (1970), further evidence has been obtained in this laboratory indicating that brain norepinephrine is affected differently when animals can avoid/escape shock from when they cannot. Weiss, Pohorecky, Dorros, Williams, Emmel, Whittlesey, and Case (1973) replicated the original findings with respect to endogenous norepinephrine levels, examining them in various regions of the brain. These results are shown in Figure 2.

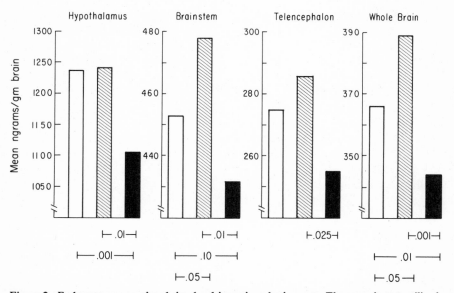

Figure 2. **Endogenous norepinephrine level in various brain areas.** The experiment utilized matched triplets (3 animals) that were run simultaneously. Each triplet consisted of an avoidance–escape animal (▧), which could control shock by its responding; a yoked animal (■), which received exactly the same shocks (via fixed tail electrodes wired in series) as did the avoidance–escape animal but which was unable to affect shock by its responding; and a non-shock control animal (□). The subjects of each triplet were exposed to their respective conditions for 20 hours prior to sacrifice for biochemical determinations. The results are based on 21 such matched triplets. At the bottom of the figure are shown the confidence levels for differences between groups.

Weiss, Pohorecky, Emmel, and Miller have also obtained some indication that animals able to control shock show higher brain norepinephrine than do animals that cannot control shock, regardless of the skeletal activity involved in performance of the coping response. For this experiment, as in earlier ones, each avoidance–escape animal again controlled shock while a matched animal (called the "yoked" animal) received exactly the same shocks as the avoidance–escape subject but could not control them. In this instance, however, half of the avoidance–escape animals (called the active condition) controlled the shock by performing a typical active behavior (running) on a running wheel, while the other half of the avoidance–escape animals (called the passive condition) were required to adopt a passive motor response (not moving) on the running wheel in order to control shock. The results of the study are shown in Figure 3. As expected, the different avoidance–escape conditions produced a very large difference in activity: in the passive condition, avoidance–escape animals showed significantly less activity than yoked animals, whereas in the active condition, they showed significantly more activity than yoked animals. However, as can be seen from Figure 4, the effect on hypothalamic norepinephrine was similar in both conditions: avoidance–escape animals showed significantly higher levels than yoked animals in both the active and the passive conditions.

Most recently, Weiss, Pohorecky, Emmel, McMeniman, Berkeley, and Jaffe estimated norepinephrine uptake and release *in vivo* by infusion of radioactive ^{3}H-norepinephrine into the ventricular system of the brain. The ventricular in-

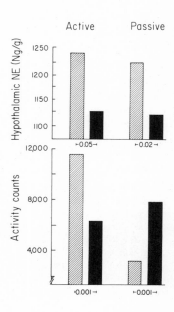

Figure 3. Mean wheel-turning activity and hypothalamic norepinephrine level for the avoidance-escape (▨) and yoked (■) animals in both the active and the passive conditions. At the bottom are shown the confidence levels for the differences between the two groups. The results are based on 24 matched avoidance–escape and yoked animals in the active condition and the same number in the passive condition. The subjects were exposed to their respective conditions for 24 hours prior to sacrifice for biochemical determinations.

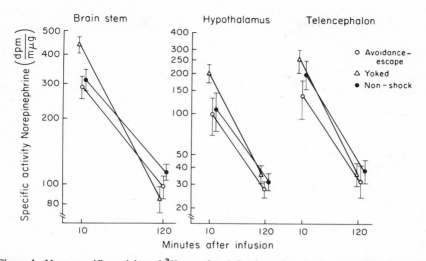

Figure 4. Mean specific activity of ^{3}H-norepinephrine in various brain areas, 10 min and 2 hours after ventricular infusion. The conditions represented by the three groups (avoidance-escape, yoked, nonshock) are explained in the legend of Figure 2. The results are based on 10 triplets at each time point.

fusion was carried out after the animals had been exposed to the experimental conditions (avoidance-escape, yoked, and nonshock control) for 20 hours. It should be noted that the infusion procedure in this study could be carried out quite quickly (less than 2 min) because the infusion was administered to unanesthetized animals through a specially developed cannula that had been implanted into the cerebral aqueduct 7–10 days prior to the experiment. In addition to speed, the use of this cannula meant that the results of the experiment were not affected by anesthetization or any operative procedure at the time of infusion, such as is generally the case with infusion studies of this kind. Immediately after the infusion, subjects were returned to their experimental conditions until sacrifice, which took place either 10 min or 2 hours later.

The results of this infusion study are shown in Figure 4. The brainstem, which was the brain region most heavily labeled by ^{3}H-norepinephrine because of its proximity to the cannula, showed the clearest effects. The rationale underlying the ^{3}H-norepinephrine infusion technique is that the ^{3}H-norepinephrine will be taken up into nerve endings following infusion and then the rate of decline of ^{3}H-norepinephrine will reflect the rate of norepinephrine release. As can be seen in Figure 4, the rate of decline (slope of line) was significantly greater in yoked animals than in avoidance-escape animals ($p < .05$) and nonshock animals ($p < .05$). Although some of the initial ^{3}H-norepinephrine uptake at 10 min was probably not specific to noradrenergic neurons and, consequently, the slopes of the curves may to some extent reflect the loss of ^{3}H-norepinephrine from non-noradrenergic sites, nevertheless the most likely explanation for these results is that the rate of release of norepinephrine was greater in yoked subjects than in the other conditions.[1] A second finding, perhaps more important, is also evident from Figure 4, namely, that the uptake of ^{3}H-norepinephrine at 10 min was greater in yoked animals than in either avoidance–escape or nonshock animals. The most likely interpretation of this particular result is that the re-uptake mechanism by which endogenous norepinephrine is normally removed from the synapse back into the nerve ending is more active in yoked animals than in avoidance–escape and nonshock animals. This interpretation is supported by the finding that adrenal steroids accelerate reuptake (Maas and Mednieks, 1971) since we have found that steroids tend to be higher in yoked animals than in avoidance–escape and nonshock animals (Weiss, 1971a).

The possibility that the reuptake mechanism is more active in yoked animals than in avoidance–escape and nonshock animals fits well with the differences in norepinephrine level that were initially found in that both types of change point to a similar behavioral consequence. Both the enhanced reuptake and the re-duced level that we have observed in yoked animals relative to avoidance–escape and nonshock animals point to the yoked animals having less norepinephrine available to postsynaptic receptors than do the animals in the other conditions. If noradrenergic activity indeed plays a significant role in mediating active motor behavior, then one could well expect yoked animals to show a deficit in such motor behavior and, hence, to show a deficit in active avoidance–escape respond-ing when compared with avoidance–escape or nonshock controls.

---

[1] It is important to note that the endogenous level of norepinephrine in these particular groups did not differ (mean level of endogenous norepinephrine in brainstem was 465.8 ± 21 ng/g for avoidance–escape subjects, 442 ± 25 for yoked subjects, and 479 ± 16 for nonshock subjects), so that differences in specific activity of ^{3}H-norepinephrine are not attributable to differences in endogenous norepinephrine. Even if one simply looks at the DPM of ^{3}H-norepinephrine in the brainstem, and thus ignores the endogenous level, the ^{3}H-norepinephrine of yoked animals is higher than that of avoidance–escape animals ($p < .01$) 10 min after infusion. The reason that the endogenous level did not differ in this particular experiment was that the subjects, instead of being housed in groups as in pre-vious experiments, were housed in individual cages to prevent them from chewing off each others' cannulas. We had previously reported (Weiss et al., 1973) that the typical depletion of brain norepinephrine level seen in yoked animals relative to avoidance-escape animals did not occur if the experimental subjects were housed individually prior to the experiment instead of in groups, so that the findings in the infusion experiment with respect to norepinephrine level were expected. It has recently been reported that individ-ual housing induces increased activity of tyrosine hydroxylase in the brain (Segal, Knapp, Kuczenski, & Mandell, 1973), the enzyme thought to be rate-limiting in the synthesis of norepinephrine, which would explain why yoked animals did not show a reduction in norepinephrine levels after having been housed individually.

RECENT BEHAVIORAL AND PHARMACOLOGICAL EXPERIMENTS

Recently, a series of 12 experiments have been carried out to further explore the possibility that the avoidance-escape deficit described by Overmier and Seligman (1967) and Seligman and Maier (1967) is indeed related to stress-induced noradrenergic changes in the brain and not to learned helplessness. It is the major purpose of the present paper to describe these experiments. Nine behavioral experiments were carried out. First, the initial experiments showed that a stressful condition (cold swim) that reduced noradrenergic activity in the brain produced an avoidance-escape deficit similar to that observed by Overmier and Seligman, while conditions that seemed to involve the same degree of "helplessness" without reducing noradrenergic activity in the brain did not produce the avoidance-escape deficit. Second, several experiments then scrutinized avoidance-escape deficits produced by both cold swim and severe, inescapable shock and showed that such deficits apparently arose from a transitory defect in the animal's ability to initiate active motor behavior, as suggested by the motor activation deficit hypothesis. Third, it is shown that *repeated* exposure to either cold swim or severe inescapable shock, which decreases noradrenergic deficiency through neurochemical "habituation," attenuates the avoidance-escape deficit, a result that is in accordance with the idea that the avoidance-escape deficit arose from stress-induced neurochemical changes. Since repeated exposure to these stressors should *increase* helplessness, this result is contrary to the "helplessness" hypothesis. Finally, three pharmacological experiments show that (1) pharmacological depletion of monoamines produces a similar avoidance-escape deficit as does exposure to severe stressors, such as cold swim and inescapable shock, and (2) subjects can be protected against effects of a severe stressor, such as inescapable shock, by drug treatments that will attenuate noradrenergic deficiency. For a more detailed exposition of these experiments, see Weiss and Glazer (1975), Weiss, Glazer, Pohorecky, Brick and Miller (1975), and Glazer, Weiss, Pohorecky, and Miller (1975).

Experiment 1. The first experiment attempted to determine whether an avoidance-escape deficit similar to that observed by Overmier and Seligman (1967) could be produced by exposing rats to a highly stressful situation that reduces brain noradrenergic activity. A forced swim in cold water (2°C) was the stressful situation used. Stone (1970a, 1970b) had demonstrated that this stressor rapidly depletes brain norepinephrine level and reduces its release; in short, the stressor apparently reduces noradrenergic activity. In addition to cold swim, animals were also exposed to swim in warm water (28°C) to test for any possible effect of helplessness that might derive from simply being exposed to a forced swim. Stone has shown that whereas swim in cold water reduces brain noradrenergic activity, the same duration of swim in warm water does not. Therefore, the extent to which a forced warm-water swim impairs avoidance–escape behavior was used to indicate the contribution of any helplessness that might derive from a forced swim and not from noradrenergic depletion.

The procedure of the experiment was as follows: two groups of animals were exposed to cold swim, of 3.5 min duration for one group and 6.5 min duration for the other. Thirty minutes later, the avoidance–escape performance of these animals was tested in a shuttle box. Two other groups received exactly the same treatments except that they swam in warm water prior to avoidance–escape testing. A fifth group received no swim before testing.

Figure 5 shows the average latency for each group to perform the correct response in a two-compartment shuttle avoidance–escape test. On each trial, a 5-sec warning signal preceded shock. If the animal jumped the 2-inch hurdle between compartments and entered the opposite compartment during this 5-sec period, it terminated the warning signal and avoided the shock. If it did not respond in this period, shock (1.0 mA intensity) was delivered and remained on until the animal responded or until 20 sec had elapsed, at which time shock was terminated. Thus, in Figure 5, latencies of less than 5 sec represent avoidance responses, latencies longer than 5 sec denote escape (from shock) responses, and a latency of 25 sec indicates a failure to escape.

The results in this first experiment showed that both the 6.5-min and the 3.5-min cold swim produced large and significant deficits in avoidance–escape

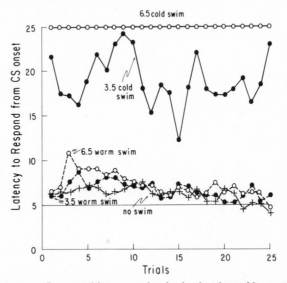

Figure 5. Mean latency (in seconds) to cross barrier in shuttle avoidance–escape test. The numbers used in designating the groups refer to the duration (in minutes) of the particular swim exposure, which was given 30 min before the shuttle avoidance–escape test. The number of subjects in each group was 6, which applies to all other experiments unless noted.

responding. The 6.5-min cold-swim animals never made an escape response. They simply lay on the grid floor of the shuttle box during the test, arching their backs and/or moving their heads when the shock began but incapable of enough motor activity to cross the barrier. The behavioral deficiency shown by these animals was more severe than any reported in the "learned helplessness" experiments and, as such, did not approximate the behavioral deficit seen in those experiments. The 3.5-min cold-swim animals, on the other hand, showed a less exaggerated deficit. These animals, while less active than the no-swim or warm-swim animals, did not lie on the grid floor and indeed could cross the barrier. When the shock occurred, these animals reared up, squealed, and moved around the side of the shuttle box on which they were standing. Nevertheless, they sometimes failed to cross the barrier at all (mean = 10.2 nonescapes), although they usually did cross before 20 sec elapsed. According to Maier et al. (1969) and Seligman et al. (1971), the characteristics by which one can identify "helplessness" in an avoidance–escape situation are (1) failures not simply to avoid shock but to escape it as well, and (2) failure of the animal to decrease its latency over trials even after escape responses are made. Both of these characteristics are evidenced by the 3.5-min cold-swim animals, so that the deficit seen in this group is a good approximation of the deficit reported originally by Seligman, Maier, and co-workers.

In that the deficit produced by a 3.5-minute cold swim is a good approximation of the deficit seen in dogs, can this deficit induced by cold swim be explained by "learned helplessness"? This seems quite unlikely. The stressful experience in this case was so brief (3.5 min) and the testing conditions (shuttle apparatus) so different from the cold-swim condition that it is hard to envision how helplessness could have been so rapidly acquired and then effectively transferred to the testing situation. Moreover, both of the warm-swim groups showed no avoidance–escape deficit, which indicates that the inescapability of the forced swim was not the factor which caused the deficit.

Thus far, evidence has been presented that (a) cold swim produces an avoidance–escape deficit similar to that observed by Seligman and co-workers, and (b) this deficit cannot be explained by learned helplessness. However, it can be suggested that some physiological effect of the cold swim treatment other than reduction of brain neurotransmitter activity was responsible for the avoidance–escape deficit. The most obvious possibility in this regard is hypothermia, which cold swim produces. There is, however, good evidence that the behavioral deficit observed in the present experiment is not due to some peripheral or nonspecific aspect of hypothermia. When Stone exposed animals to cold swim, he found, as we did in the present study, that their motor activity was reduced. The avoidance–escape deficit in the present experiment is said to derive from this deficit in the ability to elaborate active motor behavior. Stone as well as others have gathered evidence to show that this change in activity following cold swim is not due to some peripheral or nonspecific effect of the hypothermia induced by the cold swim but, rather, is produced by the reduction in central neurotransmitter activity after the cold swim. Stone showed that, after an animal had been exposed to cold swim, infusion of norepinephrine into the ventricular system of its brain increased its motor activity while *actually prolonging the hypothermia* induced by the cold swim (Stone & Mendlinger, 1974). He found that injection of amphetamine, a potent noradrenergic stimulant, also increased motor activity while exacerbating the hypothermia (Stone, 1970a). Other noradrenergic agonists, such as tricyclic antidepressants, also have this same effect on hypothermic rats (Matussek, Ruther, Ackenheil, & Giese, 1967). These results provide strong evidence that the decrease in motor activity after cold swim is not the result of peripheral or nonspecific hypothermia but is indeed the result of central neurotransmitter disruption. As a final point, it should be mentioned that Stone found that infusion of norepinephrine was more potent for increasing motor activity after swim than was infusion of dopamine, serotonin, or carbachol.

The results therefore showed that a stressor condition which markedly reduced noradrenergic activity but did not involve "learned helplessness" could produce an avoidance–escape deficit quite similar to the deficit observed by Overmier and Seligman in dogs. This is in accord with the "motor activation deficit" hypothesis. In detail, this hypothesis explains the present deficit, as well as the deficit seen in dogs, as follows. Since brain amines mediate active behavior, stressful conditions which interfere with these neurochemical systems can result in an animal being unable to produce a normal degree of active behavior under appropriate stimulus conditions. In the 3.5-min cold-swim situation (and in the original dog experiments), the neurochemical defect was such that the animal could mediate only limited attempts to escape and avoid *in a task where this level of activation was less than that required to effectively learn and consistently perform.* This emphasis on the nature of the task is important. The hypothesis holds that the animal may well be able to mediate some avoidance–escape attempts, but if the task is stressful and difficult—for example, if the animal is required to jump a hurdle into a place where it has already been shocked, as is the case with the shuttle response—it will not be able to consistently perform this response. This explains why both the dogs and the rats initially attempt to escape but then stop moving about, or occasionally perform a correct response without following up this success.

Experiment 2. The second experiment examined further the similarities between the avoidance–escape deficit induced by cold swim and that observed in the learned helplessness experiments. One observation made in the latter was that the avoidance–escape deficit could be reduced if the animals were taught the correct avoidance–escape response prior to receiving the inescapable stressor (Seligman & Maier, 1967). This second experiment showed that the deficit induced by cold swim could likewise be reduced by pretraining.

This experiment involved four groups. Two groups received exactly the same treatment as did the 3.5-min cold-swim group in Experiment 1, that is, cold swim followed 30 min later by a shuttle avoidance–escape test; and two groups received the same treatment as the no-swim group in Experiment 1, that is, no swim before the shuttle avoidance–escape test. The only difference was that one of the 3.5-min cold-swim groups and one of the no-swim groups received a single session (25 trials) of avoidance–escape training in the shuttle box 24 hours prior to the swim and avoidance–escape test. The results showed that pretraining of the correct response markedly reduced the avoidance–escape deficit following 3.5-min cold swim; the animals given pretraining prior to the cold swim and avoidance–escape test were significantly improved in the avoidance–escape test (mean latency = 9.4 sec) compared with animals not given pretraining (mean latency = 18.4 sec). Among control animals given no cold swim, pretraining had no significant effect (mean latency = 5.0 sec) compared with no pretraining (mean latency = 5.8 sec), showing that a single session of pretraining was a minor factor unless cold swim was given. Thus, the deficit induced by cold swim was similar to the deficit seen in the dogs after inescapable shock with respect to pretraining as well.

How is this effect to be explained? The explanation advanced by the proponents of learned helplessness is that giving animals experience controlling stressful events "immunizes" them against learning to be helpless because the animal thus learns that its responses are effective in altering its environment. This explanation attributes to the animal (the dog in this case) the ability to perceive that its responses are effective, and to generalize this sense of control.[2] While it may indeed be possible for a dog to learn a sense of control, the motor activation deficit hypothesis offers a simpler explanation for why pretraining ameliorates the avoidance–escape deficit. According to this view, pretraining will reduce the avoidance–escape deficit because the animal, although it has been severely stressed and therefore has limited ability to mediate motor behavior, nevertheless does not have to "waste" activity searching for the correct response but simply has to execute a response it has already learned to be correct. In short, the avoidance–escape deficit is reduced by pretraining because the

---

[2]This is probably as good a place as any to point out clearly that when one measures the effects that occur when an animal can or cannot control a stressor, this is quite different from saying that these effects occur *because* the animal is *aware* that it can or cannot control the stressor. Although our laboratory extensively studies the effects of stressor control, this refers to the specific experimental contingencies established by the experimenter and in no sense implies that the animal necessarily perceives or understands such relationships. In fact, the explanation that we have offered for effects of control makes no reference at all to the animal perceiving that it is in control or not (see Weiss, 1971a for this explanation). In contrast to this, the learned helplessness hypothesis emphasizes such cognitions (see Maier et al., 1969, p. 327).

correct response then can be accomplished quite economically with respect to motor activation.

Experiment 3. This experiment studied whether the deficit induced by cold swim would dissipate with the passage of time after the cold swim. It will be recalled that a characteristic of the deficit observed in the original "learned helplessness" experiments was that, following a single stressor session, the avoidance–escape deficit was present if the first test was given 24 hours postshock but was totally absent if the first test was given 48 hours postshock; in other words, this deficit disappeared with the passage of time. Finding a time course for the deficit induced by cold swim would thus show that this deficit was similar to that seen in the helplessness experiments in yet another respect.

This experiment was carried out as was Experiment 1, except that animals were tested in the shuttle avoidance–escape task at different intervals following the cold swim. The intervals used were 30 min, 2 hours, and 48 hours. The results showed that the effects of cold swim diminished with the passage of time. By 2 hours post-swim, effects of either the 3.5- or the 6.5-min cold swim were significantly reduced (mean latency = 8.0 and 15.8 sec, respectively) in comparison with usual effects observed 30 min postswim (mean latency = 19.8 and 25.0 sec, respectively), and by 48 hours postswim, no deficit in avoidance–escape behavior was apparent (mean latency = 6.2 and 6.0 sec, respectively). Thus, the avoidance–escape deficit following exposure to a forced cold swim, like the avoidance–escape deficit seen in dogs following a session of inescapable shock, showed a time course of decay.

Another observation made in the initial helplessness experiments was that, *once an animal had been tested in the avoidance–escape apparatus and had failed to escape*, it would then show an avoidance–escape deficit as much as 30 days after this test. To determine whether animals that initially showed an avoidance–escape deficit in the present situation would likewise show such a long-term deficit, all cold-swim groups that were initially tested within 2 hours after the cold swim were then retested 48 hours later. On retest, animals that initially showed a deficit continued to show one when retested at 48 hours.

How does one explain this long-term deficit? The proponents of the helplessness hypothesis explain both the long-term deficit and the deficit seen immediately following the stressor condition in a similar manner; namely, by proposing that the animal has learned that it can do nothing about shocks. The motor activation deficit view sees the long-term deficit, both in rats and in the dogs, as quite different in nature from the original deficit. This view states that, in the original avoidance–escape test where the animal is unable to mediate the necessary activity to accomplish the correct response, it consequently adopts other behaviors (sitting in the corner, etc.), and that such behavior is reinforced by the termination of the shock. Once this learning occurs, the animal will practice such responses in later tests. Because the competing behaviors are learned, they will mediate long-term deficits, in accord with the fact that learned responses show little dissipation with time.

Experiment 4. This experiment was the first of several which tested specific predictions from the motor activation hypothesis. If the deficit following cold swim, which appears to be analogous to the deficit produced in the learned helplessness experiments, was indeed attributable to the animal's inability to mediate sufficient active motor behavior for performance of the correct response, then raising the height of the barrier between the compartments of the shuttle box should aggravate the deficit still further. The present experiment tested this prediction by raising the barrier from its previous height of 2 inches to one of 4 inches. It was predicted that such a change would have no effect on a normal animal, thus demonstrating the sensitivity of the severely stressed subjects to the effortfulness of the task. Whereas the learned helplessness hypothesis might make a similar prediction, it is not apparent from this view that the effortfulness of the task necessarily should be a significant variable.

The experiment was carried out in the same manner as Experiment 1. Again, animals received a 3.5-min cold swim and then were tested in the avoidance-escape task 30 min later, while others received no cold swim before the avoidance–escape task. For this experiment, however, some of the animals had to jump a 4-inch barrier whereas others had to jump the normal 2-inch barrier. The results showed that animals exposed to cold swim performed significantly more poorly when the barrier height was raised to 4 inches (mean latency = 24.0 sec) from 2 inches (mean latency = 17.6 sec), and that this change had no significant effect on the performance of normal no-swim animals (mean latency, 2-inch barrier = 6.5 sec; mean latency, 4-inch barrier = 7.5 sec). These results are consistent with the motor activation deficit hypothesis.

Experiment 5. This experiment tested the converse prediction to that made in Experiment 4. If animals showed an avoidance–escape deficit in the shuttle test because they could not mediate sufficient motor behavior, then they should not show a deficit if tested in a task requiring very little motor activity. Following cold swim, animals were therefore tested in an avoidance–escape task in which the animal was placed into a tube and required simply to poke its nose through a hole at the front of it. This task, which has been used previously in our laboratory (Weiss, 1968), required only a small amount of head and neck activity to perform the correct response. This study was conducted exactly like Experiment 1, using a 3.5-min cold swim. The results are shown in Figure 6.

The results showed that, following 3.5-min cold swim, there was no deficit in learning and performance of the "nosing" response. This is in agreement with the prediction of the motor activation deficit hypothesis that responses which involve very little motor activity should be learned without difficulty.

It is also important to note that the nosing response was not simply a reflexive one; i.e., a response that occurs automatically to the shock and does not re-

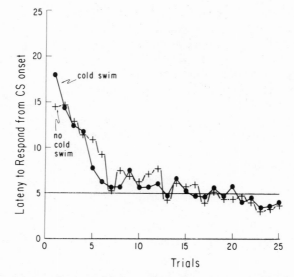

Figure 6. Mean latency (in seconds) to perform the correct response in the "nosing" avoidance–escape test. This experiment showed the effect of 3.5-min cold swim on the performance of this avoidance–escape response.

quire "voluntary" activity. Maier et al. (1973) have argued that deficits based on learned helplessness will not appear in tasks where the avoidance-escape response is reflexive because learned helplessness specifically affects voluntary response initiation. However, reflexive responses are defined as (1) short-latency responses which (2) show no learning curve (i.e., no decrease in latency over trials). It is clear from Figure 6 that this was not the case for the nosing response; this response showed initial long latencies which diminished over trials in a classical acquisition pattern. Thus, one cannot explain the similarity of the cold-swim animals and the no-swim animals on the basis that the response chosen was reflexive. Apparently, animals were able to learn a "nonreflexive" avoidance-escape response quite adequately after cold swim, providing the task involved very little motor activity.

Experiment 6. Up to this point, the experiments presented have all involved cold swim as the stressor, showing that 3.5 min of cold swim can produce an avoidance-escape deficit similar to that seen in dogs after severe inescapable shock and defining some of the characteristics of this deficit. Experiment 6 used inescapable shock to produce an avoidance-escape deficit in rats so that this deficit could be directly compared with the deficit that follows cold swim. Numerous failures have been reported in efforts to produce deficits in rats after inescapable shock (see Maier et al., 1973), and in our laboratory as well, failures have been encountered over the past few years. After manipulating many parameters, we found that the shock stressor condition had to be severe indeed. By exposing animals to shock of 4.0 mA intensity, 2.0 sec duration, occurring once every 20 sec for a total period of 45–50 min, the avoidance–escape deficit appeared. This inescapable shock was delivered through electrodes fixed on the animal's tail while the animal was confined in an apparatus shown in Weiss (1971a). When tested in the shuttle box 30 min after this stressor, the behavior of these animals was similar to that of the animals after cold swim; that is, they often did not escape within 20 sec of shock, showed variable and long latencies when they did escape (mean latency = 15.9 sec), and failed to show consistent improvement over trials. This pattern, as stated earlier, was also characteristic of the dogs in the learned helplessness experiments.

Experiment 7. Having found that a deficit in shuttle avoidance–escape responding could be produced in rats by inescapable shock, we studied whether animals exposed to this same schedule of inescapable shock would show *no* deficit in an avoidance–escape task that involved little motor activity, i.e., the nosing task. It will be recalled (from Experiment 5) that no avoidance–escape deficit was found in this nosing task after the cold-swim stressor, a result which is predicted by the motor activation deficit hypothesis but is contrary to the learned helplessness hypothesis.

Animals were exposed to the schedule of inescapable shock described above and then were tested in either the nosing or the shuttle task 30 min later. Two groups of animals that received no shock were also tested in each task. The results are shown in Figure 7. It can be seen that whereas inescapable shock clearly produced a large, significant deficit in shuttle avoidance–escape responding, the same schedule of inescapable shock produced no deficit in the nosing task. These findings therefore fit well with the view that severe inescapable shock, possibly via norepinephrine depletion, reduces the animal's ability to mediate active behavior so that (1) large deficits in responding occur in tasks requiring considerable motor activity, such as a shuttle task, but (2) small deficits or none occur in tasks requiring very little motor activity, such as the nosing task. The learned helplessness hypothesis, which holds that animals fail to avoid and escape following inescapable shock because they have learned to be "helpless," offers no reason for supposing that the deficit should depend on the motor activity required to perform the response; on the contrary, from this point of view one would expect a deficit in either task. The results therefore do not seem to be in agreement with the learned helplessness hypothesis.

Experiment 8. The next experiment tested a rather dramatic differential prediction that derives from the motor activation deficit hypothesis and the learned helplessness hypothesis. This concerns the effects of repeated exposures to the stressful situation. If poor avoidance–escape performance did in fact occur because severe stress depleted central norepinephrine and thus reduced active

motor behavior, then repeated exposure to the stressful situation should counter-
act this effect. This is because continual exposure to a stressful situation apparently causes habituation in central noradrenergic systems; that is, whereas depletion of norepinephrine may occur as the result of a single exposure to a severe
stressor, the degree of depletion is much reduced following many exposures to
the same stressor (Zigmond and Harvey, 1970). This is believed to occur because
the repeated exposure to stressors results in increased activity of the enzyme
tyrosine hydroxylase which determines the rate of norepinephrine synthesis
(Musacchio, Jalou, Kety, & Glowinski, 1969; Thoenen, 1970). In any event, if
many exposures to the stressful situation reduce norepinephrine depletion, the
motor activation deficit hypothesis predicts that many exposures to the stressful condition should actually reduce the avoidance–escape deficit in the shuttle
task. Clearly, the learned helplessness hypothesis would not predict this. Repeated exposure of the animal to a stressful situation in which it was "helpless"

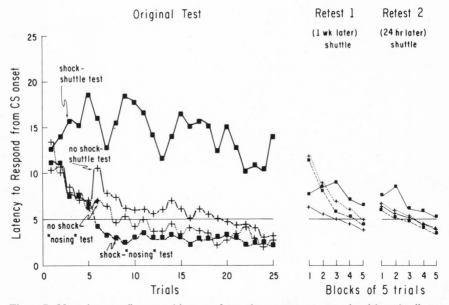

**Figure 7. Mean latency (in seconds) to perform the correct response in either the "nos-
ing" or the shuttle avoidance–escape test.** One group received inescapable shock prior to
the shuttle test (shock–shuttle test) while another group received no shock prior to this
test (no shock–shuttle test). A third group received inescapable shock prior to the nosing
avoidance–escape test (shock–nosing test) and a fourth group received no stressor prior to
the nosing test (no shock–nosing test). At the right are shown the results of two additional
shuttle avoidance–escape tests that were given to all groups, indicating that the shock–
shuttle test group continued to show a deficit in performance after they had performed
poorly in the original shuttle test.

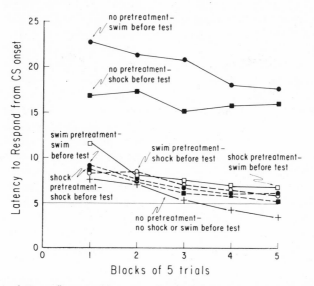

Figure 8. Mean latency (in seconds) to cross barrier in shuttle avoidance-escape test. This experiment showed the effects of repeated exposure (pretreatment) to either cold swim or inescapable shock before administration of either of these stressors and subsequent avoidance-escape test.

should, if anything, reinforce the learning of helplessness and so produce an even larger avoidance-escape deficit.

In this experiment, 4 groups of animals were given a pretreatment which consisted of exposure to inescapable shock or cold swim daily for 14 days.[3] After this pretreatment, they then received the normal stressor-test sequence used in the earlier experiments; that is, a session of either inescapable shock or cold swim followed 30 min later by a shuttle avoidance-escape test. Two groups received the same stressor 30 min before the avoidance-escape test as they had received repeatedly during the pretreatment (shock pretreatment—shock before test; swim pretreatment—swim before test), while two groups received a different stressor before the test from that received during the pretreatment (shock pretreatment—swim before test; swim pretreatment—shock before test). The latter two groups were included to test the prediction of the motor activation

---

[3]The shock sessions during the 14-day pretreatment were increased in duration over days as follows: 20 min (2 days), 25 min (2 days), 30 min (2 days), 35 min (2 days), 40 min (1 day), 45 min (1 day), 50 min (2 days), and 60 min (2 days). The session prior to the avoidance-escape test was 50 min. The swim sessions during pretreatment were also increased in duration over days as follows: 2 min (5 days), 4 min (5 days), 6 min (4 days).

deficit hypothesis that the pretreatment stressor and the stressor given just before the test need not be the same in order to be effective but that, since any
severe stressor should produce neurochemical habituation, animals could be exposed to a different stressor just before the test with the same results as if they
were exposed to the same stressor as they had previously experienced repeatedly.
Finally, three control groups were also included. First, to show the effects of
the stressors without pretreatment, two groups were simply exposed to the normal stressor-test sequence used in other experiments; that is, inescapable shock
or cold swim followed by the avoidance-escape test (no pretreatment-shock
before test; no pretreatment-swim before test). Last, the usual no-stressor control condition was also used (no pretreatment-no shock or swim before test)
to show normal performance in the avoidance-escape test.

The results, which are presented in Figure 8, showed first that repeated exposure to either cold swim or inescapable shock resulted in attenuation of the
avoidance-escape deficit in the shuttle box. Second, the results showed that this
attenuation occurred regardless of whether the stressor received before the
avoidance-escape test was the same as or different from that received during the
repeated exposures. Therefore, it appears that animals need only to repeatedly
undergo severe stress to attenuate the effect of either cold swim or inescapable
shock. These results are in agreement with the motor activation deficit hypothesis
and are contrary to expectations from the learned helplessness hypothesis.[4]

Experiment 9. This experiment examined how aspects of brain norepinephrine were affected by a number of the experimental conditions used in
the last experiment. The results of this study are shown in Figure 9. For this
experiment, four of the conditions in Experiment 8 were repeated exactly as
described above except that, instead of an avoidance-escape test, these animals

---

[4]With respect to the second conclusion, however, it should be pointed out that in the group
of animals exposed to repeated cold swim and then given a single exposure to inescapable
shock prior to the avoidance-escape test (swim pretreatment-shock before test), 3 of the 6
subjects died prior to the avoidance-escape test, so that the results for this group were
generated by the remaining 3 animals. Although these 3 animals showed no deficit in performance and were therefore distinctly different from any animal exposed to a stressor
only once, nevertheless there exists the possibility that a selection factor is represented in
the data for this group. Consequently, the findings for this group should be interpreted
with caution. However, it is equally important to note that the animals exposed to repeated sessions of shock and then given a single cold swim prior to the test (shock
pretreatment-swim before test) all survived and showed no deficit on the avoidance-
escape test. Therefore, the second conclusion mentioned above—i.e., that animals will not
show the deficit even if the stressor they receive before the test is different from that to
which they have been repeatedly exposed—is clearly supported by this experiment. [In a
later biochemical experiment (Experiment 9), one subject also died; other than this, no
subjects were lost in any of the experiments.]

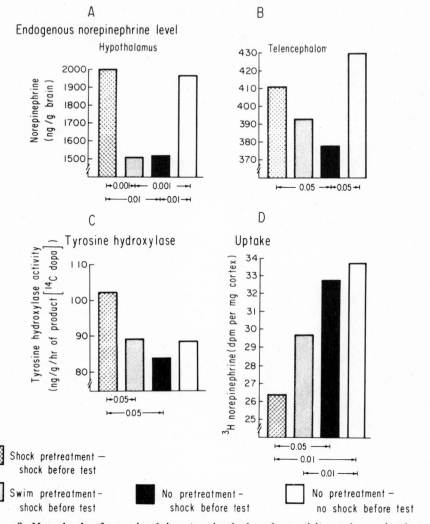

Figure 9. **Mean levels of norepinephrine, tyrosine hydroxylase activity, and norepinephrine uptake in various brain areas.** This experiment showed the effect of repeated exposure
to inescapable shock and cold swim on these measures. Two groups were repeatedly exposed to inescapable shock (shock pretreatment–shock before test) or cold swim (swim
pretreatment–swim before test), and were then sacrificed 30 min after the exposure to inescapable shock. A third group received one exposure to inescapable shock and was then
sacrificed 30 min later. A fourth group was not exposed to any stressor. The word "test"
in the key of this Figure actually stands for "sacrifice" in this particular study. The number
of subjects in each group was 12.

were sacrificed at the time when the avoidance–escape test would have been given. Brains were dissected and brain chemistry measurements were taken.[5]

First, it can be seen that a single exposure to inescapable shock (no pretreatment–shock before test) resulted in a marked depletion of brain norepinephrine concentration when contrasted with the no-shock control condition (no pretreatment–no shock before test). This depletion was seen in hypothalamic and telencephalic regions of the brain (see parts A and B of Figure 9). That the inescapable shock used produced a depletion of brain norepinephrine at the time when the avoidance–escape test normally occurred is in direct accord with the motor activation deficit hypothesis. Second, it can be seen that repeated exposure to inescapable shock (shock pretreatment–shock before test), a condition which results in no avoidance–escape deficit, produced no depletion in norepinephrine level relative to the no-shock control animals, again in accord with the motor activation deficit hypothesis. This is likely due to the induction of tyrosine hydroxylase activity seen with repeated exposure to shock (see part C of Figure 9). Perhaps most important, repeated exposure to the shock produced a large decrease in the uptake of ^{3}H-norepinephrine into brain slices (part D of Figure 9), a change which could very well mean that stimulation of postsynaptic receptors was increased because of a decrease in norepinephrine reuptake. This change in reuptake would further offset any norepinephrine depletion. Finally, it can be seen that repeated exposure to cold swim with a single shock session before the test (swim pretreatment–shock before test), a condition which also results in no avoidance–escape deficit, resulted in some reduction in norepinephrine depletion, though clearly less than was found with repeated exposure to shock. However, repeated exposure to swim did result in a significant decrease in ^{3}H-norepinephrine uptake. This indicates that repeated exposure to cold swim, like repeated exposure to shock, caused reduced reuptake, so that the increased postsynaptic receptor stimulation resulting from such a change could well have been responsible for the lack of behavioral deficit.

Further comment should be made about the "swim pretreatment–shock before test" condition. As can be seen in Figure 9, although habituation was evident in this condition, it was less pronounced with this stressor sequence than with the "shock pretreatment–shock before test" sequence. One possible explanation for this difference is that the swim–shock animals, unlike the

---

[5]Endogenous norepinephrine was separated by alumina column chromatography (Whitby, Axelrod, & Weil-Malherbe, 1961) and analyzed spectrophotofluorimetrically (von Euler & Lishajko, 1961). Tyrosine hydroxylase activity was determined by the procedure of Waymire, Bjur, and Weiner (1972). For the uptake studies, brain slices 400 μ thick were incubated for 10 min at 37°C in 5 ml of modified Krebs-Ringer's solution containing $1 \times 10^{-5}M$ pargyline and d, l-norepinephrine-7-H^3 (8.0 Ci/mmol), and saturated and agitated with 5% CO_2 in oxygen. After filtration and washing, the slices were homogenized in perchloric acid and the radioactivity in an aliquot of the supernatant was determined.

shock–shock animals, were exposed to a novel stressor before sacrifice. Although this might be correct, other evidence suggests not. In conjunction with the previous experiment, plasma corticosterone level was measured. For this measure, an extra group was included. This group was repeatedly exposed to cold swim and then exposed to cold swim (instead of shock) before sacrifice, so that this group was not exposed to a novel stressor after repeated cold swim. The results for all groups are shown in Figure 10. They show that repeated exposure to shock caused a remarkable reduction in steroid secretion when animals were exposed to shock prior to sacrifice, but no such reduction in plasma steroid level was seen in *either* group repeatedly exposed to swim (either the swim–shock or the swim–swim group). Thus, it is clear that repeated exposure to the shock had a much greater effect on steroids than did exposure to the cold swim, and this effect occurred regardless of whether the stressor before sacrifice was the same or different. This suggests that repeated exposure to the shock may have simply produced more habituation with respect to noradrenergic functioning (as well as steroids) than did exposure to cold swim.

Nevertheless, although the neurochemical and hormonal effects of repeated cold swim appear to be less pronounced than the effects of repeated exposure to shock, habituation from repeated cold swim was still evident. This is further confirmed by measures we took of brain temperature. This work was accomplished

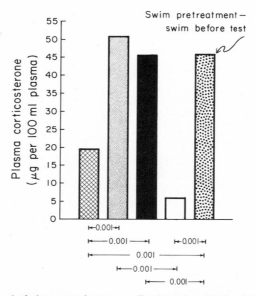

Figure 10. Mean level of plasma corticosterone. For legend pertaining to four of the groups, see Figure 9. The designation of the fifth group is shown above. The word "test" in this Figure again stands for "sacrifice" as in Figure 9.

largely through the technical facility of Mr. David Emmel, who adapted micro-thermistors so that they could be implanted directly into the brain. Recording from the thermistors implanted in the caudate nucleus, changes in brain temperature were recorded after a single exposure to cold swim of 3.5 and of 6.0 min duration, and after repeated exposures to such swim. The results are shown in Figure 11. Repeated exposure to cold swim produced no appreciable change in the rate at which brain temperature recovered after a 3.5-min cold swim, but in the case of a 6.0-min cold swim the decline in brain temperature was less pronounced and the return to normal more rapid after repeated exposure to swim.

Experiment 10. The last three experiments attempted to show that pharma-

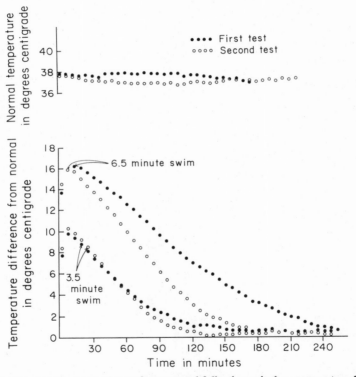

Figure 11. Mean temperature change from normal following a single exposure to cold swim and many (18-19) exposures to cold swim. Recordings were made via a thermistor implanted in the caudate nucleus. Baseline recordings were taken for all animals ($N = 6$) for 2 days to establish individual baselines across the 4 hour period examined. Animals were then exposed to a 3.5 min cold swim on one day and a 6.5 min cold swim the following day. Animals were then exposed to 15 days of cold swim of 5 min duration. After these 15 days, they were again exposed to 3.5 min and 6.5 min cold swim on successive days.

cological manipulations could affect stress-induced avoidance–escape deficits in a manner consistent with the motor activation deficit hypothesis. First, if the avoidance–escape deficit observed following inescapable shock and/or cold swim was indeed produced by a depletion of brain norepinephrine, the deficit should be reproducible by pharmacological treatment that depletes norepinephrine. In our first efforts to determine whether this prediction was correct, we administered the drug tetrabenazine, a rather nonspecific depleter of monoamines. Animals were given either the drug (2 mg/kg.) or the saline vehicle, and then were tested 30 min later. The appearance of the animals given the drug was interesting in that they were not ataxic and showed normal exploratory activity when placed on a table top; in short, on casual observation the animals appeared rather normal. However, in the shuttle box, tetrabenazine-treated animals showed an avoidance–escape deficit quite similar to that seen after cold swim or inescapable shock. On the other hand, we also tested tetrabenazine-treated animals in the nosing task described in Experiment 5, and found no deficit in this task. In other words, the results of this experiment looked very similar to those seen in Figure 7 if one simply substitutes "tetrabenazine-treated" for "shock" with respect to the two shock groups shown in the figure and "vehicle-treated" for "no shock" with respect to the two control groups. These findings showing that tetrabenazine had the same behavioral effects as inescapable shock are in accord with the motor activation deficit hypothesis.

Experiment 11. Dr. Neal Miller suggested that, just as an injection of tetrabenazine had mimicked the effects of a single inescapable shock session or a single cold swim, so repeated administration of tetrabenazine might reproduce the protective effects of repeated exposure to stressors as shown in Experiment 8 above. When tetrabenazine was repeatedly given as a pretreatment for 14 days, animals subsequently showed no deficit in shuttle avoidance–escape responding 30 min after inescapable shock or a 15th injection of tetrabenazine. Animals given placebo treatment for the same number of days showed the usual marked avoidance–escape deficit after shock or tetrabenazine. This experiment made clear that repeated depletion of monoamines, presumably resulting in compensatory habituation within these neurochemical systems, could protect animals against the effects of severe stressor and further established that the behavioral effects of stressors seen in these experiments were mediated by disturbance of such neurochemical systems.

Experiment 12. Finally, we reasoned that animals should be protected against the effects of inescapable shock if, by pharmacological means, the degradation of monoamines that the inescapable shock produced were not allowed to occur. Therefore, just before being exposed to a single session of inescapable shock, animals were given an injection of pargyline (100 mg/kg), a drug which inhibits monoamine oxidase (MAO) and would therefore prevent intraneuronal degradation of monoamines during the shock session. Such animals showed no deficit in shuttle avoidance–escape behavior 30 min after the inescapable shock session, whereas animals not given the MAO inhibitor showed the usual, large avoidance–escape deficit. Interestingly, a control group given the MAO inhibitor but no inescapable shock performed no better in the avoidance–escape task than either no-shock controls or the animals given the MAO inhibitor and the inescapable shock. This means that the MAO inhibitor did not simply balance out the effect of the shock by exciting the animal in some nonspecific way unrelated to monoamines; rather, the MAO inhibitor specifically counteracted the effect of inescapable shock. Insofar as this demonstrates that inescapable shock was having its effects via the neurochemical systems that are affected by an MAO inhibitor, it provides further evidence for the link between the stress-induced behavioral deficit and neurochemical responses that is proposed in the motor activation deficit hypothesis.

DISCUSSION

The present paper began by describing evidence which shows that a psychological variable—the ability or inability to control a stressor—can result in different changes in brain norepinephrine. Evidence was presented showing that animals that could not avoid or escape electric shocks were depleted of brain norepinephrine and had an increased reuptake of norepinephrine in comparison with animals that received the same shocks while avoiding/escaping the shock stressor. The major purpose of the present experiments, as stated in the introduction, was then to establish that these neurochemical differences could lead to behavioral differences. In this way, it would be shown how a psychological variable can alter behavior via physiological mediation that is known to us.

Toward this end, the present paper focused on a series of experiments carried out at the University of Pennsylvania beginning in the mid-1960s. The primary investigators, M. Seligman, S. Maier, and B. Overmier, had shown that animals exposed to inescapable shock subsequently could not perform an avoidance–escape response whereas animals that received the same shocks while controlling them were able to perform the avoidance–escape response. Thus these investigators had shown precisely the kind of effect required to fulfill the model in question. However, when Seligman and Maier originally reported an avoidance–escape deficit in dogs following severe inescapable shock, they attributed this deficit to the transfer of learned helplessness from the inescapable shock situation, rather than to any stress-induced physiological change. The present paper has described how certain characteristics of the avoidance–escape deficit that they reported, such as its decay over time, indicated that this learned helplessness explanation would not account for the avoidance–escape deficit, which led Weiss et al. (1970) to propose that the deficit may well have derived from a central noradrenergic deficiency produced by the inescapable shock condition. The present paper presents a series of 12 experiments to test this possibility.

The series of experiments described in this paper provide a succession of results that are consistent with the view that the avoidance–escape deficit observed by Seligman, Maier, and co-workers derived from a disturbance of central monoamines, perhaps norepinephrine. The results of 12 experiments are described in this paper. These experiments began by showing that a shuttle avoidance–escape deficit similar in key respects to that reported by Seligman, Maier, and co-workers could be produced in rats by a brief swim which reduced brain noradrenergic activity. This deficit produced by cold swim did not seem to be explainable by learned helplessness. Subsequent experiments showed that this avoidance–escape deficit (a) could be greatly reduced by pretraining of the correct response, and (b) is characterized by a time course of decay, both of these being characteristics of the deficit seen in dogs. Further experiments then showed that this deficit was (c) aggravated by raising the height of the barrier between the avoidance–escape compartments, and (d) failed to appear if the avoidance–escape task used required little motor activity, both of these findings being predicted by the motor activation deficit hypothesis.

The next part of the present investigation examined an avoidance–escape deficit produced by severe inescapable shock, studying its characteristics and comparing it to the deficit induced by cold swim. It was found that severe inescapable shock, like cold swim, produced a large deficit in shuttle avoidance–escape behavior but did not impair responding in an avoidance–escape task requiring little skeletal activity. Then, in agreement with the motor activation deficit hypothesis and contrary to the learned helplessness explanation, it was shown that a prolonged series of daily exposures to either brief cold swim or inescapable shock substantially attenuated the shuttle avoidance–escape deficit produced by either stressor. Biochemical measures were then taken which showed that a single exposure to inescapable shock depleted brain norepinephrine, a finding consistent with the presence of a deficit after one exposure, and further showed that repeated exposure to inescapable shock produced (1) no such depletion and (2) reduction in uptake, which is consistent with the absence of a deficit after many exposures.

Finally, pharmacological manipulations were introduced. It was shown that (a) a monoamine depletor, tetrabenazine, could produce an avoidance–escape deficit functionally similar to that produced by inescapable shock or cold swim, (b) chronic administration of tetrabenazine, again like the stressors, would block the deficit, and (c) pargyline, an monoamine oxidase inhibitor that should protect norepinephrine from depletion during exposure to a severe stressor, also blocked the deficit normally produced by a single session of inescapable shock.

In conclusion, it is our contention that the original phenomenon demonstrated by Overmier and Seligman (1967) and Seligman and Maier (1967) is not explained by learned helplessness but by an impairment of performance that was quite possibly produced by a disturbance of central catecholamines.

A Weak Form of Learned Helplessness?

One way to preserve the learned helplessness explanation in regard to the original dog experiments and/or the studies presented above is to postulate a weak form of learned helplessness. That is, one can concede that the avoidance-escape deficit is produced by a transient physiological change rather than the transfer of learned helplessness, but suggest that the initial physiological change itself was produced because the animal learned that it could not control the stressor in the original stress situation. Referring to the bottom portion of Figure 1, one would label the first (left) solid arrow as "learned helplessness" rather than apply this label to the dotted arrow. Such a formulation, however, contributes little to our understanding and is possibly untestable. There is, of course, no way by which one can ascertain whether a dog or any other nonverbal animal perceives its own lack of control as a prerequisite for the physiological change. As such, all that we know is that the experimental operations involved in exposing an animal to a highly stressful situation which the animal cannot control leads to physiological changes. Thus, the introduction of any notion that learned helplessness mediates the physiological change is superfluous, given our inability to test this idea.

Scope of the Motor Activation Deficit Hypothesis

It is essential to emphasize that the scope of the motor activation deficit hypothesis is much more limited than is the learned helplessness hypothesis. The learned helplessness hypothesis purports to explain not only the original dog data but a large number of experiments carried out on rats, cats, fish, chickens, and humans. In Seligman et al. (1971), for example, the number of experimental effects claimed to be explained by this hypothesis encompasses 18 published papers. The motor activation deficit hypothesis, by comparision, is claimed to explain fewer results. It certainly purports to explain all the original learned helplessness results with dogs set forth in five original research papers (Overmier & Seligman, 1967; Seligman & Maier, 1967; Seligman, Maier, & Geer, 1968; Overmier, 1968; Maier, 1970), and it may well explain certain results in other species (e.g., Richardson et al., 1970; Redmond, Maas, Dekirmenjian, & Schlemmer, 1973). On the other hand, unlike the learned helplessness hypothesis, the motor activation deficit hypothesis in no way purports to explain *all* behavioral deficits, or even all deficits in avoidance behavior which are observed following inescapable electric shock or other aversive stimuli. On the contrary, in proposing the motor activation deficit hypothesis to explain a particular set of behavioral results, the authors of the present paper are in no sense attempting to deny that considerable learning takes place during exposure to aversive and stressful conditions, and that such learning often is a powerful determinant of an animal's subsequent behavior. In fact, the study of behavioral change as produced by learning in aversive situations has long been a focus of this laboratory (e.g., Weiss et al., 1968). What is being stated in the present paper is that effects observed in a number of experiments listed above are not due to transfer of a learned response but, in this instance, are due to a transitory stress-induced neurochemical change.

It therefore should be clear from the preceding discussion that we so not rule out the possibility that some form of learned helplessness may exist, either in man or in animals. Humans can, for example, be instructed that they have no control over a stressor, and such instruction will affect physiological responses (Champion, 1950; Geer, Davison, & Gatchel, 1970). Such findings, moreover, are easily integrated into a general formulation to explain physiological effects of stress (see Weiss, 1972, p. 112). Thus, there may well be a basis for learned helplessness with respect to findings other than those discussed in this paper, particularly with human subjects.

However, despite the intuitive appeal of a learned helplessness hypothesis, it is our opinion that a considerable degree of skepticism should be maintained with respect to this type of construct. The reasons for this are threefold. First, as the foregoing sections of this paper have described, the original experiments on which the learned helplessness hypothesis were based do not seem to support the hypothesis. Second, as stated in the preceding section, it is questionable whether the existence of a cognitive "set" such as learned helplessness can be tested in a nonverbal animal. For this reason, other formulations based on directly observable variables may prove to be more fruitful in the long run both for animal and for human applications. Third, behavioral changes produced by exposure to inescapable shock conditions may be mediated, in many cases, by *learned* responses *other than* learned helplessness. We cite here one example in this category because of our familiarity with it. When Weiss et al. (1968) reported an avoidance–escape deficit following a fear-conditioning procedure, Maier et al. (1969) said that this was another demonstration of learned helplessness because the fear-conditioning procedure had involved several shocks that the animal could not control. The avoidance–escape deficit observed by Weiss et al. (1968) was indeed shown to depend on a learned response by a number of criteria. However, it was equally evident that the learned response which caused the avoidance–escape deficit was a species-specific tendency of the rat to "freeze" when afraid, and that learned helplessness was both unnecessary and unlikely as an explanation. To establish the existence of learned helplessness in the future, it will be necessary to rule out exhaustively mediation of effects by other learned responses.

Role of Norepinephrine

The motor activation deficit hypothesis has emphasized the role of norepinephrine as the principal neurotransmitter involved in the mediation of motor activity. In all the experiments where biochemical measurements were taken, indeed it was norepinephrine that was measured. We have focused our formulation on the role of norepinephrine because the hypothesis is most clearly testable in this form. On the other hand, we are well aware that norepinephrine is very probably not the only neurotransmitter involved in the central mediation of active motor behavior. For example, some recent experiments suggest that dopamine also plays a role in this respect (e.g., Thornberg & Moore, 1973). It is important to emphasize, particularly at this very early stage in our knowledge of the relationship between neurochemistry and behavior, that the motor activation deficit hypothesis is not inherently linked to norepinephrine alone. It is, rather, linked to whatever neurotransmitters are involved in the mediation of active motor behavior. As we understand more about precisely which neurotransmitters underlie active responding and how they operate to do this, the hypothesis will be capable of clearer articulation in the future.

Relevance to Human Depression

Seligman (1973) stated that lack of control in stressful situations may well be an important factor in producing "reactive" depression in humans, citing the learned helplessness experiments as a model for how lack of control results in a behavioral deficit. Insofar as such extrapolations are useful, the results presented in this paper do not contradict this view. Seligman does, of course, emphasize the *perception* of the inability to control the stressful situation, i.e., "the belief in one's own helplessness," as central to the induction and maintenance of depression. Without denying that this perception can be important, the results discussed in this paper suggest that one can maintain the importance of "lack of control" as a significant antecedent and/or concomitant of depression while emphasizing central noradrenergic deficiency as the mediator of the psychological disorder. As such, the present experiments begin to establish a link between the study of psychosocial factors that may be significant in abnormal behavior (e.g., psychological variables in stressful situations) and neurochemical hypotheses of abnormal behavior. In regard to the studies presented in this paper, the link is to the catecholamine hypothesis of depression, which states that depression is the result of a central noradrenergic deficiency. While this hypothesis almost certainly does not describe all significant neurochemical events in depression, nevertheless it can effectively organize a large body of data and has provided a promising point of departure for pursuing the link described above.

ACKNOWLEDGMENTS

We gratefully acknowledge the support of the Merrill Trust, the Sloan Foundation, Hoffmann-La Roche, FFRP (grant 72-548), and U.S. Public Health Service (grants MH 19991 and MH 13189) whose funds made the project possible. We also thank Hoffmann-La Roche for supplying the drug tetrabenazine (RO-1-9569/7).

REFERENCES

Champion, R. A. Studies of experimentally induced disturbance. *Australian Journal of Psychology*, 1950, *2*, 90–99.

Geer, J. H., Davison, G. C., & Gatchel, R. I. Reduction of stress in humans through non-veridical perceived control of aversive stimulation. *Journal of Personality and Social Psychology*, 1970, *16*, 731–738.

Glazer, H. I., Weiss, J. M., Pohorecky, L. A., & Miller, N. E. Monoamines as mediators of avoidance–escape behavior, *Psychosomatic Medicine*, 1975, *37*, 535–543.

Herman, Z. The effects of noradrenaline on rat's behaviour. *Psychopharmacologia*, 1970, *16*, 369–374.

Matussek, N., Ruther, E., Ackenheil, M., & Giese, J. Amine metabolism in the CNS during exhaustion after swimming, and the influence of antidepressants on this syndrome. In S. Garattini and M. N. G. Dukes (Eds.) *Antidepressant Drugs. Proceedings of the First International Symposium*. Amsterdam: Excerpta Medica, 1967, pp. 70–74.

Maas, J. W. & Mednieks, M. Hydrocortisone-mediated increase of norepinephrine uptake by brain slices. *Science*, 1971, *171*, 178–179.

Maier, S. F. Failure to escape traumatic shock: Incompatible skeletal motor responses or learned helplessness? *Learning and Motivation*, 1970, *1*, 157–170.

Maier, S. F., Seligman, M. E. P., & Solomon, R. L. Pavlovian fear conditioning and learned helplessness. In B. A. Campbell & R. M. Church (Eds.) *Punishment and Aversive Behavior*. New York: Appleton-Century-Crofts, 1969, pp. 299–342.

Maier, S. F., Albin, R. W., & Testa, T. J. Failure to learn to escape in rats previously exposed to inescapable shock depends on nature of escape response. *Journal of Comparative and Physiological Psychology*, 1973, *85*, 581–592.

Miller, N. E. & Weiss, J. M. Effects of somatic or visceral responses to punishment. In B. A. Campbell & R. M. Church (Eds.), *Punishment and Aversive Behavior*. New York: Appleton-Century-Crofts, 1969, pp. 343–372.

Moore, K. Effects of α-methyltyrosine on brain catecholamines and conditioned behavior in guinea pigs. *Life Sciences*, 1966, *5*, 55–65.

Musacchio, J. M., Jolou, L., Kety, S., & Glowinski, J. Increase in rat brain tyrosine hydroxylase activity produced by electroconvulsive shock. *Proceedings of the National Academy of Sciences, U.S.A.*, 1969, *63*, 1117–1119.

Overmier, J. B. Interference with avoidance behavior: Failure to avoid traumatic shock. *Journal of Experimental Psychology*, 1968, *78*, 340–343.

Overmier, J. B. & Seligman, M. E. P. Effects of inescapable shock upon subsequent escape and avoidance learning. *Journal of Comparative and Physiological Psychology*, 1967, *63*, 23–33.

Rech, R. H., Bovys, H. K., & Moore, K. Alterations in behavior and brain catecholamine levels in rats treated with α-methyltryosine. *Journal of Pharmacology and Experimental Therapeutics*, 1966, *153*, 412–419.

Redmond, D. E., Jr., Maas, J. W., Dekirmanjian, H., & Schlemmer, R. E. Changes in social behavior of monkeys after inescapable shock. *Psychosomatic Medicine*, 1973, *35*, 448.

Richardson, D., Scudder, C. L., & Karczmar. A. G. Behavioral significance of neurotransmitter changes due to environmental stress and drugs. *Pharmacologist*, 1970, *12*, 227.

Richter, C. On the phenomenon of sudden death in animals and man. *Psychosomatic Medicine*, 1957, *19*, 191–198.

Segal, D. & Mandell, A. Behavioral activation of rats during intraventricular infusion of norepinephrine. *Proceedings of the National Academy of Sciences, U.S.A.*, 1970, *66*, 289–293.

Segal, D. S., Knapp, S., Kuczenski, R. T., & Mandell, A. R. The effects of environmental isolation on behavior and regional rat brain tyrosine hydroxylase and tryptophan hydroxylase activities. *Behavioral Biology*, 1973, *8*, 47–53.

Seiden, L. S. & Peterson, D. D. Blockade of *L*-dopa reversal of reserpine-induced conditioned avoidance response suppression by disulfiram. *Journal of Pharmacology and Experimental Therapeutics*, 1968, *163*, 84–90.

Seligman, M. E. P. Fall into helplessness. *Psychology Today*, June, 1973, 7, 43–48.

Seligman, M. E. P. & Beagley, G. Learned helplessness in the rat. *Journal of Comparative and Physiological Psychology*, submitted.

Seligman, M. E. P. & Maier, S. F. Failure to escape traumatic shock. *Journal of Experimental Psychology*, 1967, *74*, 1–9.

Seligman, M. E. P., Maier, S. F., & Geer, J. The alleviation of learned helplessness in the dog. *Journal of Abnormal and Social Psychology*, 1968, *73*, 256–262.

Seligman, M. E. P., Maier, S. F., & Solomon, R. L. Unpredictable and uncontrollable aversive events. In F. R. Brush (Ed.), *Aversive Conditioning and Learning*. New York: Academic Press, 1971, pp. 347–400.

Stone, E. A. Swim-stress-induced inactivity: Relation to body temperature and brain norepinephrine, and effects of *d*-amphetamine. *Psychosomatic Medicine*, 1970a, *32*, 51–59.

Stone, E. A. Behavioral and neurochemical effects of acute swim stress are due to hypothermia. *Life Sciences*, 1970b, *9*, 877–888.

Stone, E. A. & Mendlinger, S. Effect of intraventricular amines on motor activity in hypothermic rats. *Research Communications in Chemical Pathology and Pharmacology*, 1974, *7*, 549–556.

Thoenen, H. Induction of tyrosine hydroxylase in peripheral and central adrenergic neurons by cold-exposure of rats. *Nature*, 1970, *228*, 861–862.

Thornburg, J. E. & Moore, K. E. The relative importance of dopaminergic and noradrenergic neuronal systems for the stimulation of locomotor activity induced by amphetamine and other drugs. *Neuropharmacology*, 1973, *12*, 853–866.

von Euler, U. S. & Lishajko, F. Improved technique for the fluorimetric estimation of catecholamines. *Acta Physiologica Scandinavica*, 1961, *51*, 348–355.

Waymire, J. C., Bjur, R., & Weiner, N. Assay of tyrosine hydroxylase by coupled decarboxylation of dopa formed from 1-^{14}C-*l*-tyrosine. *Analytic Biochemistry*, 1972, *43*, 588–600.

Weiss, J. M. Effects of coping responses on stress. *Journal of Comparative and Physiological Psychology*, 1968, *65*, 251–260.

Weiss, J. M. Effects of coping behavior in different warning signal conditions on stress pathology in rats. *Journal of Comparative and Physiological Psychology*, 1971a, 77, 1–13.

Weiss, J. M. Effects of punishing the coping response (conflict) on stress pathology in rats. *Journal of Comparative and Physiological Psychology*, 1971b, 77, 14–21.

Weiss, J. M. Effects of coping behavior with and without a feedback signal on stress pathology in rats. *Journal of Comparative and Physiological Psychology*, 1971c, 77, 22–30.

Weiss, J. M. Psychological factors in stress and disease. *Scientific American*, 1972, *226*(6), 104–113.

Weiss, J. M. & Glazer, H. I. Effects of acute exposure to stressors on subsequent avoidance-escape behavior. *Psychosomatic Medicine*, 1975, *37*, 499–521.

Weiss, J. M., Krieckhaus, E. E., & Conte, R. Effects of fear conditioning on subsequent avoidance behavior and movement. *Journal of Comparative and Physiological Psychology*, 1968, *65*, 413–421.

Weiss, J. M., Stone, E. A., & Harrell, N. Coping behavior and brain norepinephrine level in rats. *Journal of Comparative and Physiological Psychology*, 1970, *72*, 153–160.

Weiss, J. M., Pohorecky, L. A., Dorros, K., Williams, S., Emmel, D., Whittlesey, M., & Case, E. Coping behavior and brain norepinephrine turnover. Paper presented at Eastern Psychological Association, Washington, D.C., May 1973.

Weiss, J. M., Glazer, H. I., Pohorecky, L. A., Brick, J., & Miller, N. E. Effects of chronic exposure to stressors on avoidance–escape behavior and on brain norepinephrine. *Psychosomatic Medicine*, 1975, *37*, 522–533.

Whitby, L. G., Axelrod, J., & Weil-Malherbe, H. The fate of H^3-norepinephrine in animals. *Journal of Pharmacology and Experimental Therapeutics*, 1961, *132*, 193–201.

Zigmond, M. & Harvey, J. Resistance to central norepinephrine depletion and decreased mortality in rats chronically exposed to electric foot shock. *Journal of Neuro-Visceral Relations*, 1970, *31*, 373–381.

Discussion

MORTON F. REISER

There seemed to be two major themes in the preceding papers. The first was the old chestnut "nature versus nurture." We heard a great deal about the relative roles of constitution (genetic inheritance and prenatal and perinatal influences) and life experience in the actual development of a phenotype from the genotypic blueprint. Dr. Eibl-Eibesfeldt's paper covered this issue and talked about how ethological studies in animals can sometimes dissect out the necessary or critical contributions of one or another. His written manuscript also discussed very interestingly how cross-cultural studies can attempt to do the same thing.

Dr. Corson's paper, which is essentially an experimental study of a naturally occurring, genetically rooted animal model resembling a behavioral disorder seen in children, is an intermediate link between the more general philosophical issues touched on in Eibl-Eibesfeldt's paper and those related to the second major theme—the experimental analysis of the relationship between neurobiological and psychological parameters of behavior. Before discussing that, it should be noted at this point that there can be—and probably should be—a debate here as to whether Dr. Corson's dogs are in fact appropriately close enough models of hyperkinetic disorders in children to permit extensive analogizing between them. I'm going to deal later with what I think are the real strengths of his dog model in terms of its contribution to our understanding of clinical psychopathology. But I would be interested in further discussion of that question. After all, that issue is at the center of this volume.

The second major theme that I think Corson's work leads into very neatly is the experimental analysis of relationships between neurobiological and psy-

MORTON F. REISER • Department of Psychiatry, Yale University Medical School, New Haven, Connecticut.

chological parameters of behavior by manipulation of neurotransmitter systems
—in the case of Dr. Weiss' paper, behaviorally; and in the case of Dr. Corson's
paper, neuropharmacologically. Dr. Masserman dealt mainly with clinical behavior manipulation. (By the way, the data presented pertaining to brain-behavior
relationships happen to have considerable relevance for issues of central importance in psychosomatic medicine, and I'll try to deal with them a bit later).
Corson's work, as I noted above, is with an animal model of disease, whereas
Weiss has done a study of coping mechanisms used by animals in dealing with
experimental stress. Both address questions regarding the relation of neurotransmitters to behavior. These studies illustrate that manipulation of neurotransmitter systems provides an extremely powerful experimental strategy, and
demonstrate one of the major strengths, I think, of animal models for study of
problems in clinical psychopathology. When you study behavioral as well as
neuropharmacological experimental manipulations, data can be obtained that
will help in a number of ways and at a number of levels. First of all you can obtain data which will aid in demonstrating and elucidating the effects of environmental; that is, psychosocial, stress on the physical organism: e.g., the data
showing that the cold swim or exposure to uncontrollable shock in the rat leads
to depletion or lowering of brain norepinephrine. Second, you can obtain data
which contributes to the elucidation of the neurophysiological and neurochemical mechanisms associated with a specified behavioral state, including
altered behavioral states that are of interest to the clinician. And third, closely
related to that point, you can also obtain data that will aid in demonstrating and
elucidating the neuropsychopharmacological mechanisms underlying clinically
important effects of psychoactive drugs; for example, Corson's suggestion that
a predominently noradrenergic system would be involved in the aggressive behavior of his animals and a predominently adrenergic system in the hyperkinesis
and stereotypic behavior. The addition of selective breeding techniques to the
behavioral and neuropharmacological techniques should provide data that would
help not only with the nature–nurture problem, but also in investigating clinically, highly important questions concerning individual differences in therapeutic response to psychoactive drugs. For example, the group of wire-haired fox
terriers that are resistant to the effects of amphetamine could potentially be a
virtual gold mine in terms of helping to understand why drugs sometimes don't
work clinically in some patients. For example, there is early data which would
suggest that there may be genetically determined differences in capacity to
metabolize drugs, or in the pathways for metabolizing important psychoactive
drugs in different individuals. And this may then affect the availability and
amount of the clinically active intermediate metabolites of drugs. We have little
information clinically, for example, about the relationship of blood levels of
drugs and clinical response, and it's already clear that those alone won't be
enough because it may well be the intermediate metabolites that are the active
elements at least for some important compounds.

If one had a model system whereby you could selectively breed animals and then study the relevant enzyme systems and the metabolic pathways of drug-resistant animals, it seems to me that the field could be helped to advance a great deal further. The advantages provided by the animal model here are self-evident. One of its major strengths is its ability to provide opportunity for study of mechanisms that simply cannot be studied in man either because of the complexity of the human condition, i.e., the difficulty of controlling multivariable systems and overdetermined clinical phenomena, or because of ethical reasons. Now, it is true, that when we work with models, we're working with analogies—analogies are a form of model building. But as Samuel Butler once observed, though analogy may often be misleading, it is the *least* misleading thing we have. And so I think we ought not get caught up in too much rhetoric about absolute logical limitations.

Now, what about some of the important findings of relevance for psychosomatic medicine that I mentioned earlier? There was a phenomenon that was referred to three times in the preceding papers: the observation that the ability to cope, to control, or successfully to defend against stress-induced psychic conflict by use of psychological defenses, provides a way of protecting an organism against the potentially pathogenic physiological effects of stress. Dr. Corson mentioned it in his antidiuretic-dog model. Those animals showed the antidiuretic set of physiologic patterns, but if put into another room where they could gain mastery over the situation, they no longer displayed the reaction. And of course, the Weiss study dissects out in rather elegant detail one possible psychophysiologic mechanism whereby coping somehow enables a physiologic system to remain protected.

In this connection it should be mentioned that David Hamburg has pointed out that coping is an extremely important but under-appreciated phenomenon in human adaptation. Much of our theorizing in clinical psychopathology has always had its emphasis on disease and maladaptation, and we haven't emphasized the *positive* nearly enough. Obviously, success in adaptation is conducive to maintaining health, whereas breakdown has a great deal to do with the onset of disease. Clinically the importance of life-stress and life-crisis, in relation to the incidence of illness of all types, not of just the classic psychosomatic types, is amply proved by many studies in the literature and by the experiences of all clinicians. I just mention briefly the work of Holmes and Rahe (1967, 1969) on the relation of life-crises to the incidence of disease, and the work of Engel (1967) and Schmale (1958) which bears on clinical concepts Weiss touched on: "helplessness and hopelessness" and the "giving up–given up" complex and their relation to precipitation of disease. The frequency with which illness of any kind follows important object loss—either real, threatened or symbolic—has been documented over and over again.

There is one study in the literature that I think has not received the kind of attention that it really ought to. It is the study of Rees and Lutkins (1967) who

followed a cohort of 903 individuals for one year following death of a close relative (spouse, child, sibling, or parent). They also followed a control group in the
same community matched for age, sex, and marital status who were not bereaved. All were followed for one year—either after the loss of the loved one, or
after selection as a control. The death rate (mortality rate) in the group of
bereaved individuals in the year of bereavement was seven times that of the control group! And almost more interestingly, it was twice as high where the relative
died away from the home (including the hospital) as it was when death occurred
at home. So that when we talk about separation (see the preceding papers on
separation and anaclitic depression in the animal), and when we talk about life-
crisis and ability to cope and so forth, we are talking about phenomena that
are important, not only in terms of contributing to or protecting against
psychological difficulties in living, but to actual maintenance of life itself.
It seems to me these phenomena bear centrally in the mystery of mind-body
relationships.

These data imply that if you can cope with stress or loss adequately, you can
avoid physiologic turbulence and consequences and be less susceptible to successful invasion by pathogenic factors, be they infectious, neoplastic, climatic, or
whatever. They fit, also, with some other experimental data, for example, the
classic study of Sachar and his colleagues at Walter Reed Hospital (1963) that
showed reciprocity between effectiveness of psychological defenses and level of
activation of the pituitary–adrenal cortical axis in patients with acute schizophrenia. In acute panic states they displayed two- to four-fold increases in
urinary 17-hydroxy corticosteroid, epinephrine, and norepinephrine levels. When
defenses were put back together—even if they were psychotic defenses—the
levels of adrenal hormones came back to normal. Similar observations on this
reciprocity has been reported many times and in chronic states as well, (Friedman, Mason, & Hamburg, 1963). I think I ought not to leave the impression—it
would be a false one—that the relationship between coping or defense and physiologic change is as simple as the data presented suggests—even the data that I've
just mentioned. On the surface, the data suggest a linear relationship such that
when defenses are working there is virtually no physiological arousal—and when
they break down, you get more and more physiologic response. However, I
would wonder if the intensity of stress—and by that I mean the strain on the
organism, i.e., the degree to which it breaks up homeostasis—may not be an important factor. For example, is there a threshold phenomenon in this? If stress
gets great enough, will even perfect defenses fail to protect "breakthrough" into
the physiologic system? For example the very strenuous avoidance procedures
used by Mason, Brody, and Tolliver (1968) on monkeys who were coping very
well none-the-less produced considerable adrenal cortical arousal in those animals. Or if you had very mild stress, would you, even with breakdown of defenses, see little or no physiological activation? Then there would also be the

question of whether you might be dealing with a U curve or an inverted U curve as you go from virtually little strain to maximal strain. These are all parameters of these critically important phenomena that still need to be investigated. There are a couple of other points to mention in respect to the human in whom it gets even more complicated. Bourne, Rose, and Mason (1968), in a military situation, have shown that it's not only the quality of the stress and effectiveness of defenses that are important, but also uncertainty, group support, and the individuals' roles in the social system. For example, in a group of men in a combat zone in South Viet Nam waiting for attack, the officers had very marked physiological arousal, and the enlisted men had very little or none. Any number of parameters that influence how this defensive system works still need to be explicated.

I just want to add one point about Dr. Weiss's paper. It is terribly frustrating to look at data from only one system. What would his data look like if he had taken an endocrine profile—adrenal medullary and cortical hormones, thyroid, growth hormone, and so on—would they move together or would they move in different directions? What would be the sequential relationships? We know that the steroids, for example, influence the rate of synthesis and breakdown of the monoamines in the important neurotransmitter systems. We also know that the neurotransmitters have something to do with activating the release of some of the endocrines; at least we know that dopamine apparently is the substance responsible for the ultimate release of growth hormone from the anterior pituitary. And Dr. Strobel will probably say that you have to look also at the question of biologic rhythms in experiments of that kind. And I'm sure that Seymour Levine would say, in looking at any of these experiments that if you had manipulated the early developmental history of the animals, you might have influenced the nature of these relationships as displayed by the adult rats. I mention these things just to provoke some discussion of Dr. Weiss's paper as I promised to do. I hasten to add that I think his experiments are truly elegant. Seldom have I seen a more satisfying or pleasing series of experiments, exercises in logic, and experimental strategy and tactics, but I have a hunch it would be almost too lucky to be true to happen upon the final answer at first try looking only at a single substance, brain norepinephrine. I think—for a variety of reasons (that I'm sure are going to come up later in discussions)—that there's a great deal that still needs to be done to make the findings convincing, and I think that the teleological explanation is a little premature.

I don't want to stop without saying that I hope Dr. Eibl-Eibesfeldt's paper will come in for considerable discussion. I found it absolutely fascinating. His concept—that man has, by his capacity to learn and store knowledge, another route for adapting to the environmental challenge in an evolutionary way and even the chance to overcome the problem of war, for example say, by deliberate and controlled application of this capacity—is in my view an extremely important contribution.

REFERENCES

Bourne, P. G., Rose, R. M., & Mason, J. W. 17-OHCS levels in contact special forces "A" team under threat of attack. *A.M.A. Archives of General Psychiatry*, 1968, 135–140.

Engel, G. L. A psychological setting of somatic disease: The "giving up-given up" complex. *Proceedings of the Royal Society of Medicine*, 1967, *60*, 553

Friedman, S. B., Mason, J. W., & Hamburg, D. A. Urinary 17-hydroxycorticosteroid levels in parents of children with neoplastic disease. *Psychosomatic Medicine*, 1963, *25*, 364–376.

Holmes, T. H. & Rahe, R. H. The social readjustment scale. *Journal of Psychosomatic Research*, 1967, *11*, 213.

Mason, J. W., Brody, J. V., & Tolliver, G. A. Plasma and urinary 17-hydroxycorticosteroid responses to 72-hour avoidance sessions in the monkey. *Psychosomatic Medicine*, 1968, *30*, 608–630.

Rahe, R. H. In P. R. A. May, J. R. Wittenborn (Eds.), *Life crisis and health change, psychotropic drug response: Advances in prediction*. Springfield: Charles C. Thomas, 1969, p. 92.

Rees, W. D., & Lutkins, S. G. Mortality and bereavement. *British Medical Journal*, 1967, *4*, 13–16.

Sachar, E. J., Mason, J., Kolmer, H. S., & Artiss, K. L. Psychoendocrine aspects of acute schizophrenic reactions. *Psychosomatic Medicine*, 1963, *25*, 510–537.

Schmale, A. H. A relationship of separation and depression to disease. *Psychosomatic Medicine*, 1958, *20*, 259.

WORKSHOP II: Conflict of Adaptation to Changed or Induced Environmental Conditions

Edited by HOWARD F. HUNT

The wide-ranging discussion centered around three major topics: the paradoxically beneficial, calming effects of amphetamine on hyperkinetic, disturbed dogs as a model for understanding similar effects of stimulant and antidepressant drugs on hyperkinetic children; animal models for aggression in relation to war and crimes of violence in man; and the contribution of animal models to understanding of human psychopathology. Rather than attempt a recapitulation in dialogue form, this report will simply paraphrase and condense major comments related to these topics to provide continuity and organization.

In explanation of his amphetamine findings, Dr. Corson suggested that both hyperkinetic dogs and hyperkinetic children have defects in inhibitory systems. These defects presumably are based on lack or deficiency of certain neurotransmitters, probably adrenergic, in inhibitory brain areas. Within a narrow, moderate dose range, amphetamines either replace missing transmitters or, more likely, restore a proper balance between excitation and inhibition, permitting selective attention and improved regulation of behavioral outputs in the face of distractions. In large doses, amphetamine excites hyperkinetic dogs just as lower doses affect normal dogs. Chemical analysis to verify hypothesized anomalies in con-

Howard F. Hunt • New York State Psychiatric Institute, Columbia University, New York, New York. (Workshop moderated by Howard F. Hunt.)

centration or turnover of neurotransmitters in various areas of the brain in hyperkinetic dogs remains to be done, however.

Dr. Samuel Irwin suggested that it might be the ability of low doses of the stimulant drugs to lower threshold for arousal that is important, enabling the hyperkinetic organism to achieve levels of arousal optimum for effective functioning. He saw a parallel with arousal-enhancing, risk-taking activities among sociopaths who are thought to have high thresholds for auditory stimulation (as measured by evoked potentials in the brain). These people, and perhaps hyperkinetic individuals, need a lot more stimulus input than normal persons to feel organized, functional, and normal, and engage in their symptomatic behavior in attempts to generate such inputs. Studies of evoked potentials in the brains of hyperkinetic individuals could shed light on this possibility.

Sharp controversy developed around the adequacy of animal models for human aggression. The innateness of the aggression seen in the animal models themselves, the identity of these behaviors across situations and across species, and the biological role of aggression in natural selection were hotly debated, without resolution of the problems. Dr. Robert Hinde doubted the usefulness of analogies, suggesting that behaviors with similar form or similar consequences may have very different biological bases. Further, he questioned Lorenz's energy model of aggression. Finally, he suggested that the notion of species "preprogramming" presented by Dr. Eibl-Eibesfeldt oversimplified matters if the origins of adaptive behavior represent a different problem from that of the results of subtle interplay between organism and environment in development.

Dr. Eibl-Eibesfeldt recast the concept as "potentiality in development" ("in the blueprint of the species"), with precise mechanisms obscure because we know so little about the embryology of behavior. He further stressed that he was not invoking an energy model of aggression, but rather arguing that aggressive behavior in animals is in the service of the important biological function of spacing, among other things, and that phylogenetic adaptation determines its course. Motor responses involved in aggression are in the motor repertoire of the species as "fixed action patterns" that may appear in individuals never before exposed to conspecifics. Because slaying conspecifics has questionable value for species survival, such phylogenetic adaptations tend to be ritualized and thus to spare the antagonist. In man, however, he saw the use of weapons in war and other destructive aggression as a cultural invention. Here, submissive postures or actions cannot terminate the destruction, as in the animal model, because they are not observed in time or are not observable at all, given the physical separation of the adversaries. In the animal model, and perhaps in human interpersonal aggression without major weapons of war, actions characteristic of submission can terminate the aggression. War and interpersonal aggression in man functions as an inbuilt motive (according to the energy model); Dr. Eibl-Eibesfeldt felt that the possibility should not be summarily discarded as foolish. Rather, the matter deserves further study.

Drs. L. J. West and Seymour Levine both spoke against the loose and confusing use of "aggression" and "violence." Like "anxiety," these terms are applied to such diverse matters that they lose their meaning: e.g., outcomes for the victim or target; motives, instincts, or drives; modes of striving or ways of going about satisfying other needs. These uses refer to very different biological and psychological situations. To what, then, are the animal models to be analogous? Dr. West joined Dr. Eibl-Eibesfeldt in questioning territoriality among animals as the basis for destructive aggression in man. To be sure, man conducts wars against outgroups, but in man, "going to war" can be seen as conformity to outside social and governmental pressure, and passivity in following a line of least resistance for the individual, rather than a burgeoning of instinctive aggression. Indeed, Dr. West noted that the majority of crimes of violence, such as murders, are visited upon the criminal's near or dear, not upon outgroup members. He saw such behavior as a matter to be understood as arising out of the person's psychodynamics, with animal models of behavior to be used for furthering our understanding of such specifics, not directed toward discovering some ethologically determined innate drive toward destructive aggression.

Similarly, Dr. Levine emphasized the role of specific conditioning and early experience in producing animal behaviors that, to an observer, might appear aggressive and violent but that actually served other motivational masters. Dr. Hunt pointed out that such behavior even could arise adjunctively (like forced drinking or pica in the rat) as a result of too lean schedules of reward for other behavior. Dr. West and Dr. Jules Masserman pointed to a number of animal preparations characterized by disturbed behavior, based on conditioning or special experiences, that could serve as models for human disorders, but that do not require assumptions about innate drives for explanation (e.g., Masserman's "experimental neuroses," Harlow and Suomi's monkeys, Brady's "executive monkey"). They are created in the laboratory, as prototypes, and even methods of therapy can be tried out on them for suggestions as to what might work for people. Dr. West went on to suggest that there may be naturally occurring characteristics having survival value for animals that, when enhanced in force or concentrated in frequency, may be prototypes of biologically colored human disorders such as schizophrenia (e.g., musth in the male Indian elephant).

In a plea for animal models of greater relevance to the clinician, Dr. Shein suggested the desirability of developing genetic strains of animals showing particular susceptibility or resistance to behavioral pathology. Then, study of these strains, and differences between them, might shed light on specific mechanisms producing the pathology, to aid in the development of rational clinical treatment. Drs. Kling and Strobel commented on the existence of several such strains, and on work in progress along these lines. Dr. Levine, however, emphasized most forcefully the almost insuperable difficulties in separating the effects of "nature" (innate biological inheritance) from those of "nurture" (the effects of environment and experience). Development represents a process that includes the interaction of the two, not just the summation of their separate contributions, and the process starts at conception not at birth. Dr. Kolb noted that the parents of battered children tend, statistically, to have been battered themselves in childhood. Here, a clearly experiential effect is transmitted from generation to generation, to appear in each successive generation after a latency of 20–25 years. Do animal models exist for this phenomenon?

The adequacy and usefulness of animal models for human behavior pathology was questioned at a number of points throughout the meeting. These models tend to be analogies, and some of the participants saw the value of analogies to be limited as explanatory devices, however attractive they might be as illustrations. Both Dr. Meltzer and Dr. Hunt stressed the high priority that should be given to studying mechanisms involved in the pathologies of interest, on doing experiments that are worthwhile for themselves and not just because they represent attractive analogies of superficial relevance to practical problems. The meeting closed with a cautionary note to the effect that while a good model may help us think constructively about a problem, help us to move scientifically in the right direction, a bad model actually may foreclose progress. As an example, the "learned helplessness" model is analogically related to depression and defects in coping; the relevance to depression is strengthened by Dr. Weiss' findings on norepinephrine in relation to "helplessness." But to make "learned helplessness" *the* model for depression because it seems so relevant at this stage of knowledge would be to press too hard. After all, the psychological aspects of the animal model give an essentially cognitive anwer to a behavioral problem in species not generally characterized by extensive cognitive behavior. Answers to the problem of depression based on similarities between the model and depression thus may be nonanswers. Also, by partially satisfying curiosity, these answers might foreclose more basic research endeavors on psychological and biological mechanisms directly related to depression in man.

Neurophysiological Experimental Modification of the Animal Model as Applied to Man

The Use of Differences and Similarities in Comparative Psychopathology

R. A. HINDE

INTRODUCTION

In any symposium concerned with the relevance of animal data to man, a biologist is inevitably ambivalent. Since his science is unified by the theory of evolution by natural selection, he must hold that there are or have been continuities between animals and man—a view that is reinforced as more and more similarities at the molecular and biochemical levels are revealed. But at the same time, much of a biologist's training is concerned with the nature of species differences. He has learned how even closely related species are adapted to different enviornments, and how the resulting differences are superimposed on broad phyletic changes of a more fundamental nature, including differences in complexity of organization (e.g., Schneirla, 1949). The awareness of adaptation and consequent diversity has caused him to have reservations about the psychologist's search for a comprehensive theory of behavior (Hinde & Stevenson-Hinde, 1973; Seligman & Hager, 1972), and to doubt the simplistic belief that the capacities of higher organisms can be understood directly in terms of more elementary functions (Schneirla, 1949). Thus, a biologist's training promotes awareness both of the continuities and of the differences between animals and man.

In discussing the use of animals for research in psychopathology, should we

R. A. HINDE ● MRC Unit on the Development and Integration of Behaviour, University Sub-Department of Animal Behaviour, Madingley, Cambridge, England.

focus on the similarities of animals to man or on their differences from him? The phrases "the animal psychopathological model" in the title of this symposium, and "validity of animal models for man" in the subtitle, suggest that it is the similarities on which we should focus. The word model implies a resemblance; and models are usually evaluated in terms of their similarity to the original. But the value of models in science lies in part in their differences from the original. They are useful because, by virtue of availability or simplicity, they pose questions, suggest relations or can be manipulated in ways not possible with the original. When a model becomes an exact replica, it loses its *raison d'être* (Arber, 1954; Craik, 1943; Hinde, 1956; Toulmin, 1953).

Here, of course, lies a dilemma. The closer a model to the original, the more likely are the questions it raises to be pertinent and the relations it suggests to be valid also for the original. But if it resembles the original too closely, its value as a model diminishes, and one might as well work with the original. There is also another danger: as the model approaches the original it becomes increasingly easy for the investigator to confuse the two, to assume that all the properties of the model exist also in the original, and to confuse the two in arguments in which the model is employed.

The point of view put forward here is this. A search for particular animal models for use in studying particular problems in psychopathology, in the belief that the more similar the model to man the more relevant the data, is useful in some instances, but is fraught with dangers. Since animal models must be both similar to and different from man, it is often valuable to focus on the differences as well as on the similarities, and to exploit both. Furthermore, to search for principles is often more likely to be fertile than to look merely for similarities in overt behavior.

Animal models are used in psychopathology for two main reasons. One is ethical: in our society it is permissible *for certain purposes* to experiment with animals; to rear them under conditions of deprivation; to irradiate, ablate, inject or innoculate them in a way that would be quite improper with man. There are differences even in the extent to which mere observation of animals and of man is permissible. To observe the copulatory behavior of birds, or to follow consort pairs of chimpanzees seeking seclusion in the forest, are meritorious occupations: analogous activities with human subjects are, or were until recently, questionable even in scientists of impeccable integrity. In general, doubts about the propriety of observing, manipulating or experimenting with animals decrease the more distant they are from man. The converse of this must not be forgotten—in studying psychopathology, ethical considerations become more important, the closer the model to man. The use of any species must be justified in terms of the problem under study, and higher mammals, primates, and especially apes, should never be used if lower species will do.

The second reason for employing an animal model is scientific. Animals are simpler, and perhaps less variable, than man. They can be reared under controlled conditions and still passed off as reasonably normal. Their behavior can be measured with moderate precision. As we have seen, on scientific grounds the optimum model is one close to man, but not too close. It is not possible to generalize about the most desirable degree of difference: this will vary with the problem in hand, and with the characteristics of the various possible models that are relevant to the problem. For some purposes even the chimpanzee, the animal probably closest to man in most respects, is not close enough. For others, a more distantly related primate might be desirable, and for yet others the best model may be found outside the primate order.

SOME DANGERS IN FOCUSING ON SIMILARITIES

If we choose an animal model solely on the basis of its similarities to man, there are certain dangers which must be borne in mind. First, the very act of comparing may lead us to underestimate the complexity of the human case. The finding of overt similarities between certain aspects of the behavior of animal and man does not mean that concepts or principles derived from the study of animals are wholly adequate to account for the behavior of man. The point has been forcibly made by Rosenblatt (1974a) in the context of pathological aggression. Studies of animals have been of value in permitting the isolation of many of the factors conducive to aggression, such as proximity, male sex hormone, and frustration. However, this does not necessarily mean that we can fully understand human aggression on the basis of its animal counterparts. Rosenblatt cites an (admittedly fictitious) example of a man who murdered a woman whose sexual attractiveness might have made him leave an environment in which he felt himself to be secure, pointing out that to describe such an incident in the terms we use to describe animal behavior neglects the complexity and deviousness with which humans deal with feelings and emotions.

Furthermore, in comparing animal with man it must not be forgotten that scientific methodology itself tends to give us an over-simple view of the behavior of animals. Even in the initial stages of describing behavior we abstract regularities and ignore apparent irregularities—irregularities here being observations that lie outside our preconceived or nascent conception of how things are. We can test the ability of apes to solve problems, but can make generalizations only about what we find they can do, not about what other individuals in other circumstances might do. We can ascribe emotions or cognitions to them on the basis of what we can interpret of what we observe, but not of what we do not

see or cannot understand. Thus, while some may be tempted to anthropo-
morphize, the more hard-headed the scientist the more is he likely to over-
simplify. This is important here because the application of concepts derived
from the study of relatively simple animals to the immeasurably more complex
human case is rendered the more hazardous if those concepts tend to simplify
even the animal case from the start. Analogizing may lead us into even more
gross simplifications in our attempts to understand man.

A second danger arises from the fact that behavior has multiple determinants. Let us consider one experiment involving a species considerably lower
down the scale than man—the laboratory rat—and relatively simple types of behavior, such as those shown in a strange arena or open field. The experiment was
designed to investigate how these were affected by various types of treatment in
infancy (Denenberg & Whimbey, 1968; Denenberg, 1970). Four aspects of the
animals' early experience were varied:

1. Some of the mothers of the experimental animals were "handled" in their
infancy while others were not.

2. Some of the experimental animals were themselves handled in infancy,
while others were not.

3. Some were born and reared to weaning in an ordinary laboratory maternity cage while others were reared in a larger "free environment" cage of greater
complexity.

4. After weaning some animals were kept in small laboratory cages, and
some in "free environment" cages.

The experimental design is shown in Table I. With the additional variable of
sex it resulted in 32 groups, and 3 animals were tested in each.

The animals were tested in an open field in adulthood, and a variety of measures taken. Of these, some were strongly intercorrelated and could be labeled as
"emotional reactivity," another group could be characterized as "consumption-
elimination," and a third as "field exploration." Table II shows how these were
affected by the various treatment factors. It depends on an analysis of variance
of a variety of measures of behavior in diverse situations. Only one aspect of the
results is pertinent here. Emotional reactivity was affected by early handling of
the pups and, since the analysis of variance revealed only one interaction, this effect of handling was relatively invariant over the other experimental manipulations used in the study. By contrast, exploratory behavior was affected (though
in opposite directions) by handling of both the experimental animals themselves
in infancy and by handling their mothers in infancy, and there were seven
significant interactions. In other words, the influence of each of these main
determinants was affected by the other factors manipulated. This is exemplified
in Table III, which shows the effects of handling and the pre- and postweaning
environments on activity. The possible combinations of treatments are ranked in
the order of the activity scores that resulted. Clearly, handling has an effect—the

Table I. Experimental Design Used by Denenberg (1970).[a]

Conception to day 21 experience of natural mothers in their infancy	Days 1–20 handling experience of pups	Days 1–21 preweaning housing	Days 21–42 postweaning housing
NH	NH	MC	LC
NH	NH	MC	FE
NH	NH	FE	LC
NH	NH	FE	FE
NH	H	MC	LC
NH	H	MC	FE
NH	H	FE	LC
NH	H	FE	FE
H	NH	MC	LC
H	NH	MC	FE
H	NH	FE	LC
H	NH	FE	FE
H	H	MC	LC
H	H	MC	FE
H	H	FE	LC
H	H	FE	FE

[a]NH = nonhandled control, H = handled, MC = maternity cage, FE = free environment, LC = laboratory cage.

Table II. Qualitative Summary of the Results of an Analysis of Variance of the Experiment shown in Table I (From Denenberg, 1970)

Factor variable	Significantly modified by	Effect
Emotional reactivity	Handling pups in infancy (5 of 5)[a], one interaction	Reduces emotionality
Exploratory behavior	Handling pups in infancy (3 of 5)	Increases exploratory behavior
	Handling mothers during their infancy (3 of 5), seven interactions	Decreases exploratory behavior
Consumption– elimination	Handling pups in infancy (1 of 3)	Decreases defecation
	Handling mothers during their infancy (1 of 3)	Decreases food intake
	Postweaning housing in free environment (1 of 3) Two interactions	Decreases water intake

[a]Numbers in parentheses indicate number of significant effects out of total number possible. For example, five tests defined the factor of emotional reactivity; in all instances the offspring handling main effect was found to be significant.

Table III. Open-Field Activity: Means for the Interaction of
Offspring Handling × Preweaning Housing × Postweaning
Housing (Denenberg, 1968)

Handling	Preweaning	Postweaning	Mean[a]
No	Free environment	Free environment	7.75
No	Cage	Cage	12.58
Yes	Cage	Free environment	17.08
No	Free environment	Cage	29.83
No	Cage	Free enviornment	42.08
Yes	Free environment	Cage	51.75
Yes	Free environment	Free environment	72.08
Yes	Cage	Cage	74.25

[a]Number of squares entered.

two least active groups were not handled and the two most active were. Given that the rats are either handled or not handled and the environment is held constant through weaning, the nature of that environment seems to have a relatively small effect on activity. But if the environment is changed at weaning, the effect depends on the sequence of environments experienced: free environment followed by laboratory cage enhances exploratory activity more than the opposite sequence. Furthermore, even this effect is modified by handling experience.

Here, I believe, is an important lesson for comparative studies of psychopathology. The experiment shows that a given type of experience in infancy may affect a given type of adult behavior hardly at all, moderately, or dramatically, depending on the other experiences of the subject. Thus simple generalizations about the effects of a particular type of early experience are almost inevitably misleading. If this is true within one species, how much more must it be true when we attempt to generalize from one species to another. If a certain type of early experience is conducive to adult pathology in one species, we must be extremely wary in generalizing that finding to another, however closely related.

In this context the patterns taken by most comparative studies of behavior pathology must be noted. One common practice is to select a particular human symptom or syndrome, such as depression, to find an animal model in which similar symptoms can be induced, and to study the etiology of those symptoms in the animal. Another procedure is to start at the other end of the causal sequence, selecting an independent variable with known pathological consequences in man, such as crowding or isolation, and studying its consequences in an animal model (Figure 1). In either case it is necessary to remember that pathological syndromes have multiple causes and involve a complex nexus of events: we may be able to identify one point of the nexus in the animal model with one point of the nexus in man, but as we move away from that point, similarity becomes increasingly less probable. Thus total social isolation in a

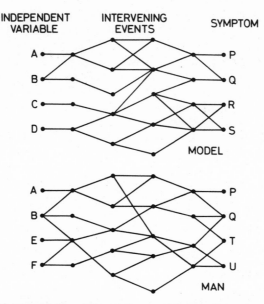

Figure 1. **Schematic view of the relations between etiology and symptoms in animal model and in man.** In this hypothetical case two independent variables (A + B) and two symptoms (P + Q) are similar in both. The immediate consequences of A + B, and the immediate antecedents of P + Q, are also similar but not identical. The intervening nexus differs. In most animal models the similarities to man lie in either the independent or the dependent variables, but rarely in both.

vertical chamber is an effective way of inducing depression in rhesus monkeys, but this is certainly not the way in which depression is usually caused in man. However, that fact does not in itself mean that such experiments are unjustifiable: if the symptoms of depression in monkey and man really are similar, the closer within the causal nexus that we come to those symptoms, the more likely are the nexuses to correspond. In other words, even if the psychological or behavioral antecedents of depression are dissimilar between monkey and man, the biochemical intermediaries may yet be similar (McKinney, Suomi, & Harlow, 1971). Similarly, if we start from the independent variable, the ultimate symptoms may differ markedly between species even though the mechanisms are similar. Thus although post mortem examinations of animals subject to severe crowding show lethal lesions differing markedly between species, the hypothalamic–adrenocortical system may be an intermediary in them all (e.g., Calhoun, 1962; Henry, Meeham, & Stephens, 1967; von Holst, 1972).

The third danger in drawing analogies from animal to man is closely related to the last, and was cogently emphasized by the late Dr. Lehrman (1974). It concerns the temptation to select particular processes, or particular aspects of behavior, and to compare them in isolation from the total pattern of life of each species. Lehrman discussed especially the case of temporary separations between mother and infant. Over 20 years ago it was suggested that such separations may have adverse effects on behavior development (Bowlby, 1951, 1969; Spitz, 1946), but for a long time the matter was controversial, and the important variables are still far from clear (e.g., Rutter, 1972; Douglas, 1975). A number of investigators therefore used rhesus monkeys as models, investigating the effects of periods of separation from the mother on their behavior after reunion (e.g., Seay & Harlow, 1965; Hinde & Spencer-Booth, 1971). The symptoms found, involving "protest," "despair," and "depression," were clearly analogous to some of those shown by human children. Evidence was also forthcoming that the effects of even a short separation experience could be long-lasting (Spencer-Booth & Hinde, 1971). This suggests that long-term effects may occur also in man. It was further found that, when mothers are temporarily removed from their infants, the distress exhibited by the infants after reunion varies with the nature of the mother–infant relationship before separation: those infants whose attempts to gain ventro–ventral contact had been rejected least by their mothers, and who had had to play the least role in maintaining proximity with her when off her, tended to be least distressed after reunion (Hinde & Spencer-Booth, 1971). Those characteristics of the mother–infant relationship that mitigate against postreunion distress in rhesus monkeys seem analogous to those found to be important in the human mother–child relationship (Schaffer and Emerson, 1964). On these and other grounds there seem to be considerable parallels between the consequences of a separation experience in young rhesus and those in human children.

Turning now to comparisons with closely related species, Kaufman and Rosenblum (1969) and Rosenblum (1971) have carried out comparable experiments with two other species of macaque, the bonnet and the pigtail monkeys. While the pigtail infants responded in much the same way as the rhesus, the bonnets did not show depression to anything like the same extent. This is perhaps related to a difference in social structure between the species: whereas, during the mothers' absence, the rhesus and pigtail infants were largely ignored by the other adults present, the bonnet infants received much attention from other females. Thus, data about the effects of separation in one species cannot necessarily be generalized to another species, even though that species is congeneric.

Furthermore, even in rhesus the nature of the mother–infant relationship may be affected by the social situation in which mother and infant live: the relationship tends to be looser in mother–infant pairs living alone in large cages than in similarly caged groups (Hinde & Spencer-Booth, 1967). Since the consequences of a separation experience vary with the mother–infant relationship, they may also be indirectly affected by the social environment. Thus, generalizations about the effects of a separation experience must be made with caution not only across species, but also when they concern mother–infant pairs in different environments.

In discussing these results, Lehrman (1974) pointed out that the relevance of the monkey data for man might have appeared in a different light had the early experiments been conducted with bonnet macaques rather than rhesus. A short separation experience in infancy produces immediate distress and detectable long-term effects in socially living rhesus monkeys, but not in bonnets kept under comparable conditions. Thus, attempts to generalize from a particular animal model to man may well lead to false conclusions. Lehrman also emphasized that human societies differ greatly from each other, and that the effects of a period of mother–infant separation may well not be the same in cultures with an extended family structure as they are in western Europe or the United States. Indeed, patterns of mother–child interaction vary even between the social classes within any one society (Bernal & Richards, 1973). If the possible animal models and the social milieux within the human species are both diverse, what chances are there that analogies from a particular model for a particular society will be useful? At this point I have put the issue pessimistically: I shall return to it later.

FOCUSING ON DIFFERENCES

If there are such dangers in comparative studies, are they really any use for the human case? Are animal models ever likely to be useful in human psychopathology? I believe that they are, but to be so they must be used in a way which exploits not only their resemblances to man, but also their differences from him. Some of the issues here are obvious and have already been mentioned —the ethical issues that require the use of nonhuman species, and the scientific convenience for many purposes of animals that differ from man in that they can be bred under controlled conditions and experimented upon. There are, however, some points that deserve emphasis.

First, it often happens that particular aspects of behavior are especially well developed or accessible for study in some nonhuman species. Study of such a species may call attention to their importance or permit their investigation in a manner not possible in man. Since early mother–infant relations are thought often to be important in the development of psychopathology, we may consider some examples of the ways in which the study of simpler species has been revealing for the human case.

1. *Contact Comfort.* The contact comfort that infant monkeys obtain from their mothers plays an important part in their development (Harlow & Zimmermann, 1959): its study led to a better appreciation of the importance of contact

comfort for the human baby. But this appreciation came about partly just because contact comfort is *more* important in monkey than in man.

2. *Orientation to mother.* The young primate comes to direct a wide variety
of responses to its mother. Studies of mammals in which the young are kept for
a while in a nest, such as the domestic cat, reveal that orientation to the mother
and to the nest develop with considerable independence from each other. Rosenblatt (1974) points out an inherent problem in studying mother–infant development in primates is that, just because the mother is such a central figure, it may
be impossibly difficult to disentangle processes which in other species are directed toward different objects and thus develop separately.

3. *Relative roles of mother and infant.* Perhaps because they can tell
mothers what to do when babies pay them no respect, pediatricians have been
wont to emphasize the extent to which parents control the behavior of their
infants and to disregard the ways in which infants control their parents. A more
balanced view has long been prevalent in the animal literature (Harper, 1970),
and its relevance to man is now apparent (Bell, 1968).

4. *Weaning.* In considering the growth of independence, some psychologists
and pediatricians have emphasized the child's growing ability to free himself
from his possessive mother (Rheingold & Eckerman, 1971). Study of the much
simpler situation in animals has permitted some conceptual clarification in this
area by demonstrating the necessity for separating the questions:

a. How does the nature of the mother–infant interaction determine each
aspect of the relationship at each age?

b. What are the nature of the changes in the mother–infant interaction that
determine the changes in the relationship with age? Do changes in the mother or
changes in the infant determine the age changes in the relationship?

c. Are differences between mothers or differences between infants primarily
responsible for differences between mother–infant pairs at any one age?

To consider just one aspect of the relationship, the spatial proximity of
mother and infant, studies of rhesus monkeys indicate that in that species the
mother is primarily responsible for maintenance of proximity in the early weeks,
the infant later; changes in the mother are primarily responsible for changes in
proximity with age; and the relative importance of differences between infants
and of differences between mothers changes in a complex way with age (Hinde,
1969; Hinde & White, 1974). The several questions must thus be considered
independently of each other.

These examples have been chosen because of their diversity—they show some
of the different ways in which comparative studies can assist understanding of an
important problem in human psychopathology: in each case they are helpful in
part because animals are different from man. Just in case it should be thought
that the argument is confined to early development, consider two aspects of

the control of aggression. While observational evidence strongly suggests that "frustration" plays a crucial role in human aggression (Dollard, Doob, Miller, Mowrer, & Sears, 1939; Miller, 1959), experiments with human subjects are in danger of being unethical, and their artificiality sometimes makes them unconvincing (but see Berkowitz, 1974): experiments with animals permit generalizations about the role of frustration in animals (e.g., Azrin, Hutchinson, & Hake, 1966) whose applicability to man can then be assessed. Similarly the effect of punishment on aggression is inevitably complex, since pain can augment aggression and thus counter the effects of the punishment. The effects are difficult to disentangle in man and the problem is so important that a search for principles by animal experimentation is justified (Ulrich & Symannek, 1969). In such cases it is just because animals are simpler than man that their use enables theoretical issues to be highlighted that would otherwise be shrouded in the complexity of the human case.

But in every one of these cases the dangers of comparative studies must be constantly borne in mind. The attractiveness of the mother rhesus for her infant is not due only to the contact comfort that she provides: surrogate mothers who are warm, provide milk, or rock, are more attractive than those who lack these characteristics (Harlow & Suomi, 1970). The infant's responses to his mother are multifactorially determined, and the relative importance of those factors must be expected to differ between species. It is fruitless to search for the cause for aggression: an aggressive episode depends upon multiple predisposing and precipitating factors whose relative importance will vary from incident to incident, individual to individual, and even more from species to species (Hinde, 1974). Again, no one should think of applying findings from rats about the effect of experience in infancy on activity in an open field in adulthood, or from rhesus monkeys about the role of contact comfort or the effect of mother–infant separation, directly to man, without considering first how the vast differences in sensory and motor abilities, in the nature of the mother–infant relationship and in the social environment, may affect each case.

Mention must be made of one other way in which species differences can be valuable. If we can produce a particular symptom in a relatively simple animal, and account for it in terms of concepts applicable to that animal, then this limits the complexity of the concepts we *need* invoke to account for it in man. If separation induces depression in monkeys, then it must be explicable in terms of concepts applicable to the monkey level of cognitive functioning. If separation-induced depression in man is identical with that in monkey, then it *may need* no more complex concepts for its explanation. Of course the words "may need" must be emphasized here: the etiology may be quite different between monkey and man, and the dangers of oversimplifying the human case, discussed earlier, must be borne in mind. But knowledge that simpler concepts may do can help us to economize in our theorizing.

ABSTRACTING PRINCIPLES

This brings me to my final point, which in a way summarizes what has gone before. The emphasis in the term "animal psychopathological model" suggests that for a specific human problem we search for a specific animal model on which we can perform specific manipulations that will produce symptoms comparable to those we wish to study. This is one way of proceeding, and it works in some cases: we have seen examples in the study of depression, of mother–infant relations, and of aggression. We have also considered some of the dangers that beset such enterprises: what we must beware of, at all costs, is focusing merely on superficial similarities in overt behavior. We have also seen that it is in part because animal models are different from man that they are useful, and that it is often profitable to focus on the differences as well as on the similarities. This leads to another way of using the animal data—namely attempting not to find superficial similarities between animal and man, but to abstract principles. This involves the study of both differences and similarities, and can provide an in-depth understanding of the phenomena under study whose applicability to the human case is more readily assessed.

Let me first take one example. Pediatricians have been in almost constant disagreement about the relative merits of schedule and demand feeding. Considerable light is shed on this controversy by consideration of the relation between milk composition and suckling frequency in other mammals. In general, species that suckle their young seldom have milk with a high protein content, while species that suckle their young almost continuously have less concentrated milk. Human milk has a relatively low protein concentration (Blurton Jones, 1972). In this relatively concrete case a principle concerning the relationship between milk composition and suckling frequency can suggest conclusions about the "natural" (and what that means is another question not to be discussed here) suckling frequency in man.

Another more complex case has already been mentioned. As Lehrman (1974) pointed out, the conclusions to be drawn from studies of mother–infant separation in macaques will depend on the species chosen: bonnet macaques appear to be much less affected by a separation experience than pigtails or rhesus, the difference probably being related to differences in their social structure. Thus there are dangers in the use of a particular animal model, and these are magnified by the diversity of human societies. But if we use not one animal model but several, and if in each case we study the various factors that affect the response to separation, we can then abstract principles that may find application to humans. We can in fact use both the differences within species caused by interaction with the social environment, and the differences between the species, to give a much deeper understanding of the separation syndrome and its etiology.

In such cases, a problem of crucial importance concerns the level of abstraction or analysis at which principles are to be sought. I know of no guidelines here, except that generality of a principle tends to be correlated with superficiality, and the one must be weighed against the other. However one further example from the study of mother–infant separation illustrates the importance of the issue. In our early experiments we achieved separation by removing the mother from the pen for a week or two, leaving the infant in the physical environment and with the social companions to which he was used. In this way the infant's separation experience was not contaminated by exposure to a strange environment. Later we reversed the procedure, removing the infant to a strange place and leaving the mother in the home pen. We expected that these infants, having had a more traumatic experience, would show more distress on reunion than the first group. Exactly the opposite was the case. The infants whose separation had involved removal to a strange place showed less distress after reunion than those whose mothers had been removed. At the data level, this seemed to be contrary to the human case: although no hard information exists, it seems probable that children who go away to a strange place for a while are subsequently more distressed than children whose mothers go, leaving them at home. But when we came to examine the behavior of the separated monkeys in more detail it became apparent that the infants who were themselves removed were reunited with mothers who had suffered relatively little trauma, and who could thus meet their temporarily increased demands for contact and maternal solicitude. By contrast, the mothers who had themselves been removed had to recover from the trauma of their own removal and re-establish their relationships with group companions. They thus had less to spare for their infants, who therefore continued to show considerable distress after reunion (Hinde and Davies, 1972). In other words, distress after reunion depends on the quality of the mother–infant relationship, and thus in turn on the whole social situation. This incidentally illustrates again the importance of examining the whole situation in depth, and not merely isolating the fragment in which one is initially interested. But it also suggests that it is profitable to compare animal and human not at the data level, but at a level at which principles can be abstracted. At the data level, comparison of baby-removed and mother-removed separations show opposite tendencies in monkey and man. But the principles, that postreunion distress depends on the quality of the mother–infant relationship, and this on the whole social situation, are likely to be common to both.

Thus, principles are more likely to be useful the more species or situations on which they are based. This may sound like a truism, but it is too often neglected in psychopathological studies that look for a point-to-point resemblance between animal model and man. And of course it contains another danger; namely, animals are so diverse that it is easy to select facts to fit theories and to neglect awkward cases. The probable relevance of principles derived from a spectrum of animal models must be assessed with reference to their phylogenetic, social, and ecological similarities to and differences from man.

In this paper I have been concerned specifically with the question of animal models in psychopathology. This is only one of the areas in which the disciplined use of comparative data can help in the understanding of human behavior, but it would not be relevant to discuss here the ways in which studies of nonhuman species can further knowledge of specific aspects of behavior, suggest questions, provide principles, permit clarification of conceptual issues, or give perspective to man in emphasizing both his evolutionary heritage and his uniqueness. Over many issues, comparative data are more likely to be relevant, the nearer their sources to man. But it is not only in psychopathology that animal models are of most value if they are used not merely for comparison of isolated fragments of behavior and their etiology and consequences between animal and man, but for the abstraction of principles. If those principles are based on a conscious awareness of multifactorial determination and on a knowledge of the total social situation and biology of the species, then their applicability to man in all his social diversity is likely to be fruitful.

REFERENCES

Arber, A. The Mind and the Eye. Cambridge: The University Press, 1954, pp. 40–44.

Azrin, N. H., Hutchinson, R. R., & Hake, D. F. Extinction-induced agression. *J. Exp. Anal. Behav.*, *9*, 1966, 191–204.

Bell, R. Q. A reinterpretation of the direction of effects in studies of socialization. *Psychol. Rev.*, 1968, *75*, 81–95.

Berkowitz, L. In Aggression—Origins and Determinants, W. W. Hartup & J. de Wit (Eds.). The Hague: Mouton, 1974.

Bernal, J. F., & Richards, M. P. M. What can zoologists tell us about human development? In S. A. Barnett (Ed.), Ethology and Development. London: Heinemann, 1973.

Blurton-Jones, N. G. (Ed.) Ethological studies of child behaviour, Cambridge: The University Press, 1972.

Bowlby, J. Maternal Care and Mental Health. Geneva: World Health Organization, 1951.

Bowlby, J. Attachment and Loss (Vol. 1). London: Hogarth, 1969.

Calhoun, J. B. Population density and social pathology. *Scientific American*, 1962, *206*, 139–148.

Craik, K. J. W. The Nature of Explanation. Cambridge: The University Press, 1943.

Denenberg, V. H. Experimental programming of life histories and the creation of individual differences: a review. In M. R. Jones (Ed.), Miami Symposium on the Prediction of Behaviour, 1968: Effects of Early Experience, Miami: University of Miami Press, 1970, pp. 61–91.

Denenberg, V. H., & Whimbey, A. E. Experimental programming of life histories: toward an experimental science of individual differences. *Developmental Psychobiology*, 1968 *1*(1), 55–59.

Dollard, J., Doob, L. W. Miller, N. E. Mowrer, O. H., & Sears, R. R. Frustration and Aggression. New Haven: Yale University Press, 1939.

Douglas, J. W. B. Early hospital admissions and later disturbances of behaviour and learning. *Developmental Medicine and Child Neurology*, 1975, *17*, 456–480.

Harlow, H. F., & Suomi, S. J. Nature of love–simplified, *Amer. Psychologist*, 1970, *25*, 161–168.

Harlow, H. F., & Zimmermann, R. R. Affectional responses in the infant monkey. *Science*, 1959, *130*, 421–432.

Harper, L. V. Ontogenetic and phylogenetic functions of the parent-offspring relationship in mammals. *Adv. Study Behav.*, 1970, *3*, 75–119.

Henry, J. P., Meehan, J. P., & Stephens, P. M. The use of psychosocial stimuli to induce prolonged systolic hypertension in mice. *Psychosom. Med.*, 1967, *29*, 408–433.

Hinde, R. A. Ethological models and the concept of "drive." *The British Journal for the Philosophy of Science*, 1956, *VI*, 24, 321–331.

Hinde, R. A. Analyzing the roles of the partners in a behavioural interaction–mother-infant relations in rhesus macaques. *Ann. N.Y. Acad. Sci.*, 1969, *159*, 651–667.

Hinde, R. A. The study of aggression–determinants, consequences, goals and functions. In W. W. Hartup & J. De Wit (Eds.), Aggression, Origins and Determinants. The Hague: Mouton, 1974.

Hinde, R. A. & Davies, L. M. Changes in mother–infant relationship after separation in rhesus monkeys. *Nature,·*1972, *239*, 41–42.

Hinde, R. A., & Spencer-Booth, Y. The effect of social companions on mother-infant relations in rhesus monkeys. In D. Morris (Ed.), Primate Ethology. London: Weidenfeld and Nicolson, 1967.

Hinde, R. A., & Spencer-Booth, Y. Effects of brief separation from mother on rhesus monkeys. *Science*, 1971, *173*, 111–118.

Hinde, R. A., & Stevenson-Hinde, J. (Eds.). Constraints on Learning: Limitations and Predispositions. London: Academic Press, 1973.

Hinde, R. A., & White, L. The dynamics of a relationship–rhesus monkey ventro-ventral contact. *J. Comp Physiol. Psychol. 86*, 8–23.

Holst, D. von. Renal failure as the cause of death in *Tupaia belangeri* exposed to persistent social stress. *J. Comp. Physiol.*, 1972, *78*, 286–306.

Kaufman, I. C., & Rosenblum, L. A. Effects of separation from mother on the emotional behavior of infant monkeys. *Ann. N.Y. Acad. Sci.*, 1969, *159*, 681–695.

Lehrman, D. S. In N. F. White (Ed.), Ethology and Psychiatry. Toronto: University of Toronto Press, 1974.

McKinney, W. T., Suomi, S. J., & Harlow, H. F. Depression in primates. *Amer. J. Psychiatry*, 1971, *127*, 1313–1320.

Miller, N. E. Liberalization of basic S-R concepts. In S. Koch (Ed.), Psychology a Study of a Science, Study 1, Vol. 2, New York: McGraw-Hill, 1959.

Rheingold, H. L., & Eckerman, C. O. Departures from the mother. In H. R. Schaffer (Ed.), The Origins of Human Social Relations. London: Academic Press, 1971.

Rosenblatt, D. S. In N. F. White (Ed.), Ethology and Psychiatry, Toronto: University of Toronto Press, 1974.

Rosenblum, L. A. Infant attachment in monkeys. In H. R. Schaffer (Ed.), The Origins of Human Social Relations. London: Academic Press, 1971.

Rutter, M. Maternal Deprivation. Harmonsworth: Penguin, 1972.

Schaffer, H. R., & Emerson, P. E. The development of social attachments in infancy. *Monogr. Soc. Res. Child Devel.*, 1964, *29*, 3.

Schneirla, T. C. Levels in the psychological capacities of animals. In R. W. Sellars et al. (Ed.), Philosophy for the Future. New York: Macmillan, 1949.

Seay, B., & Harlow, H. F. Maternal separation in the rhesus monkey. *J. Nerv. Ment. Dis.*, 1965, *140*, 434–441.

Seligman, M. E. P., & Hager, J. L. (Eds.) Biological Boundaries of Learning. New York: Appleton-Century-Crofts, 1972.

Spencer-Booth, Y. S., & Hinde, R. A. Effects of brief separations from mothers during infancy on behaviour of rhesus monkeys 6–24 months later. *J. Child Psychol. Psychiat.*, 1971, *12*, 157–172.

Spitz, R. A. Anaclitic depression. *Psychoanal. Study Child.*, 1946, *2*, 313–342.

Toulmin, S. E. The Philosophy of Science. London: Hutchinson, 1953, pp. 31-39, 165–167.

Ulrich, R., & Symannek, B. Pain as a stimulus for aggression. In S. Garattini & E. B. Sigg (Eds.), Aggressive Behaviour. Amsterdam: Excerpta Medica, 1969.

Animal Models for Brain Research

JOSÉ M. R. DELGADO

Many biological mechanisms investigated in animals yield results applicable to man. Enzymatic functions are shared by microorganisms as well as by human beings. Respiratory properties of hemoglobin are similar in mice, whales, and man. Basic knowledge of nerve physiology has been obtained from frogs and squid.

The human brain also has many functions which have been investigated in elemental models including the octopus, bees, and man-made machines such as computers. Some aspects of behavior, however, are unique to man and it is therefore difficult, if not impossible, to investigate in animals the neurological basis of human culture. It is doubtful whether animal models can be found for schizophrenia and other mental disturbances, whereas sensory perceptions have been successfully studied in rats, cats, and dogs.

The model to be chosen depends obviously on the experimental theme: brain chemistry, requiring the use of a large number of animals, is usually performed in inexpensive subjects such as rodents; for higher intellectual functions such as abstract thinking and speech, the great apes have been tested.

During the last three decades, technical developments have made possible direct communication with the depth of the brain in both animals and man, circumventing physiological receptors and effectors. For this research, tiny sensors and probes are implanted within the brain tissue and connected to ex-

JOSÉ M. R. DELGADO • National Center "Ramón y Cajal" (Seguridad Social) and Autonomous Medical School, Madrid, Spain.

ternal instruments to obtain electrical and chemical information, or to deliver messages and drugs. Implantations have been made in different species, including rats, cats, monkeys, and chimpanzees. In addition, very fine electrodes have been placed within the brains of patients for diagnostic and therapeutic purposes.

The field of brain research is expanding very rapidly and in the last few years new advances have made possible the establishment of radio links between brain and instruments, two-way communication between brain and computers, and transdermal stimulation of the human brain. In this paper we present a brief outline of available methodology and some speculations about expected developments. Radio communication with the brain of freely moving subjects represents a new experimental approach for the understanding and possible control of psychological functions in both animals and man.

TECHNIQUES FOR ELECTRICAL STUDIES

Most techniques for electrical exploration of the brain are based on the introduction within the cerebral tissue of fine metallic conductors, protected with insulated material except at the tips, terminating in small sockets located outside the scalp. Connections with instrumentation for stimulation or recording are easily established by attaching suitable leads.

Recent developments include:

1. Modern stereotaxic surgery which permits the safe and accurate placement of a large number of electrodes within specific brain targets. This technique is well known and is used in many laboratories around the world (see review in Sheer, 1961).

2. A miniaturized, multichannel radio stimulator mounted on a small harness permits electrical brain stimulation (ESB) of the completely unrestrained subject. The device developed by our laboratory (Delgado, 1963, 1969a) for use in both animals and man, consists of two instruments:

(a) The Radio Frequency (RF) transmitter, which measures 30 × 25 × 15 cm and includes the necessary electronics for controlling the repetition rate, duration, and intensity of the stimulating pulses. The intensity of ESB is controlled by varying the frequency of the subcarrier oscillators. Transmission is performed in the 100 MHz band.

(b) The receiver stimulator, small enough to be carried by the subject, measures 37 × 30 × 14 mm and weighs 20 g. The instrument has solid-state circuitry encapsulated in epoxy resin to make it waterproof and animal proof. As the output intensity is related to the frequency of the subcarriers, it is independent of changes of strength in the received signal, making the instrument highly reliable.

The output is of constant current and therefore independent of wide changes in biological impedance.

3. Telemetry of electrical signals has been used by many investigators (see review in Delgado, 1963, 1970). In general, a miniature amplifier–FM transmitter combination and a telemetry receiver are used for this purpose. The transmitting unit used in our laboratory consists of an EEG amplifier with a gain of 100, input impedance of 2 MΩ, frequency response of from 2 to 200 Hz and a voltage-controlled oscillator for each channel. The outputs of the subcarrier oscillators are summed and connected to a single RF transmitter operating at 216 MHz. The electrical signals from the intracerebral electrodes are received and magnified by the amplifier. The output of this amplifier controls the frequency of the sub-carrier oscillator, and the oscillator output in turn controls the frequency of the transmitter. With this instrumentation, intracerebral correlates of conditioning, learning, decision making and other behavioral events may be investigated by remote control.

4. The integration of several channels for radio stimulation of the brain and for telemetry of depth electroencephalography (EEG) constitutes the "stimo-ceiver" (*stimu*lator and EEG re*ceiver*). This unit is small enough to be carried by a monkey on a collar, or by a patient under a head bandage. The instrument per-mits continuous recording of spontaneous electrical activity of the brain, and also makes it possible to send programmed stimulations by radio while the sub-jects are completely free, as explained later.

5. With a stimoceiver, brain-to-computer-to-brain radio communication has been established in the chimpanzee. In these experiments, spontaneous bursts of spindles in both amygdalas were telemetered, recorded and recognized by an on-line computer which automatically sent radio stimulations to a negative rein-forcing area of the reticular formation each time that a spindle was identified by the computer. After 2 hours of this contingent stimulation, ipsilateral spindling was reduced 50%. After 6 days of contingent stimulation for 2 hours daily, spindling was reduced to only 1%, and the chimpanzee had diminished attention and motivation. These effects lasted for about two weeks and were both revers-ible and reproducible. Results from this study demonstrated:

(a) The feasibility of direct communication from brain to computer to brain, circumventing sensory receptors.

(b) The suppression of a specific EEG pattern by contingent radio stimula-tion of a specific intracerebral point.

(c) The possibility of learning by means of direct electrical stimulation of a brain area (for more details, see Delgado, Johnston, Wallace, & Bradley, 1970).

6. The terminal socket of intracerebral electrode assemblies, by which the depth of the brain is made accessible either by direct connection of leads to laboratory instruments or to a radio stimulator, has always been a visible artifact and potential source of infection. To avoid these problems, we developed a mi-

crominiaturized, batteryless stimulator 26 mm in diameter and 8 mm thick, which can be implanted permanently underneath the skin. Its terminal leads are placed stereotaxically within the brain. Power and information are transmitted through the intact skin to the subcutaneous instrument by means of radio induction. Pulse duration, frequency, intensity, and duration of stimulation can be controlled remotely for four cerebral points. The system can be compared to cardiac pacemakers, although it has much greater electronic complexity, and it has been referred to as a brain pacemaker. It has been used successfully in the monkey, gibbon, and chimpanzee, and recently in man. Some experiments have continued for over a year with excellent tolerance, no changes in local excitability, and reliable effects of stimulation (Delgado, Rivera, & Mir, 1971; also see section E). Transdermal electrical stimulation of the brain permits the application of programmed excitation of selected cerebral structures for as long as needed, providing new possibilities for the therapy of brain disturbances.

7. We have recently developed a new generation of stimoceivers, not for external use as before, but for totally subcutaneous insertion, permitting several channels of communication from as well as to the brain, through the intact skin. This two-way transdermal unit at present is being tested in animals.

In the near future, we should be able to process electrical information collected transdermally from different cerebral structures and, with the aid of on-line computers, to correlate it with spontaneous and evoked behavior. We should also be able to influence the brain directly, sending programmed, contingent, automatic radio stimulation to specific cerebral structures, for example, to avoid the occurrence of undesirable electrical patterns related to the onset of epileptic or aggressive episodes. The establishment of artificial intracerebral links and biofeedbacks could thus enhance, inhibit, or otherwise modify specific brain functions.

These predicted developments should provide new possibilities for diagnosis and therapy based on the immediate and automatic exchange of information between brains and computers.

RESULTS OF ELECTRICAL STUDIES

1. At the unicellular level, it is known that there is anatomical and functional specificity for relatively simple visual inputs. For example, as demonstrated by Hubel and Wiesel (1962), in the visual cortex, some cells respond to simple stimuli such as bars and edges, while other neurons respond to the rate and direction of movement. It is doubtful, however, that perceptual and psychological activities could be correlated with discharges of single neurons. We know almost nothing about coding for the transmission and evaluation of information. Inves-

tigation of these questions (the unravelling of intracerebral codes of communication) is one of the main challenges of brain research.

2. At the multicellular level, electrical changes have been detected in specific areas of the brain during learning, conditioning, instrumental responses, and other activities. For example, bursts of theta rhythm have been recorded during an animal's approach to food cued by a light (Grastyán, Karmos, Vereczkey, Martin, & Kellenyi, 1965). In cats, different hippocampal EEG activity has been detected when the animals were alert and quiet, during orienting reactions, or while performing discriminative responses (see bibliography in Jasper & Smirnov, 1960 and Quarton, Melnechuk, & Schmitt, 1967). Computer analysis of EEG may help to identify the anatomical structures involved and the specific electrical patterns related to determined types of behavior. We do not know, however, the meanings of the recorded waves or whether or not they are essential for behavioral performance.

3. In animals, a variety of autonomic functions have been driven or modified by ESB. For example, the diameter of the pupils can be precisely controlled by adjusting the intensity of ESB applied to the lateral hypothalamus. This response can be maintained indefinitely in monkeys, and it represents the introduction of an artificial functional bias, changing the set point of normal pupillary reactivity to light (Delgado & Mir, 1966). A wide range of behavioral manifestations including motor activity, food intake, aggression, maternal relations, sexual activity, motivation, and learning have been influenced by ESB (see examples in Figures 1 and 2, and summary and references in Delgado, 1969b).

4. An important limitation of ESB is that it is merely a nonspecific trigger of preestablished brain functions; ESB is incapable of providing information comparable to that acquired through the senses, and it cannot induce any precisely predetermined behavior. Stimulation of points in the central gray area may induce aggressive behavior in a monkey, but details of the animal's performance will be in agreement with its previous experience and present environmental circumstance. Rage may be induced electrically, but its expression and direction cannot be controlled. If an aversive stimulation is applied to an animal when it is dominant in a group, it may attack a submissive member with which it has had previous unfriendly relations; the same stimulation applied when the subject is in an inferior hierarchical position may evoke only a submissive grimace.

All experimental evidence indicates that it is highly unlikely that the details of behavioral performance can be directed by ESB, and it is even more improbable that a robot-like activity could be induced in animals or man by electricity.

5. The existence of inhibitory mechanisms in the central nervous system was reported a century ago by Sechenov, and the representation of inhibition within specific brain structures has been widely explored (Beritoff, 1965; Diamond, Balvin, & Diamond, 1963; Pavlov, 1957; see also discussion by Delgado, 1964).

Studies, mainly in animals, have revealed that stimulation of points within the septum, caudate nucleus, amygdala, thalamus, and reticular formation may produce the following effects: (a) motor inhibition, including adynamia, arrest reaction, and local muscular paralysis; (b) inhibition of food intake; (c) inhibition of aggression; (d) inhibition of maternal behavior; (e) autonomic inhibition, including slowing down of the heart, decreased respiration, and lowering of blood pressure; (f) sleep.

The arrest reaction is a remarkable effect consisting of the sudden cessation of spontaneous behavior; it is as if a motion picture projector has been stopped, freezing the ongoing action. A cat arrested while lapping milk stayed with its tongue out and another cat, stimulated while climbing stairs, froze between two steps.

Aggressive behavior of the normally ferocious rhesus monkey can be inhibited by stimulation of the head of the caudate nucleus, and the artificially pacified animal can be petted and even touched on the mouth safely. Animals as dangerous as brave bulls have also been arrested in full charge by radio stimu-

Figure 1. Radio stimulation of the central gray area in the large gibbon induces aggressive behavior, directed against the other gibbon, which reacts with suitable offensive-defensive behavior.

Figure 2. Radio stimulation of the large gibbon in the head of the caudate nucleus inhibits the animal which remains motionless throughout the period of stimulation. Observe on the other gibbon the harness to carry a transmitter of mobility, housed inside a plastic cylinder.

lation of the caudate nucleus. In addition to these results of brief duration, programmed stimulation of inhibitory areas can induce prolonged effects; for example, intermittent radio stimulation of the caudate nucleus in the boss of a monkey colony may abolish its dominance and change the hierarchical structure of the whole group.

Several types of cerebral disturbances in man, including hypermotility, anxiety, and epilepsy, are perhaps related to an intermittent, excessive neuronal activity of brain structures that could be identified. If undesirable intracerebral activity, related to behavioral episodes detrimental to a human subject, could be inhibited by programmed ESB, the therapeutic consequences would be most beneficial.

At present, much less is known about brain inhibition than brain excitation, but it is certain that both processes are closely correlated and requisite for the performance of behavior. Every act requires the selection of a specific pattern from the many available. In order to think, we must choose one subject and suppress a continuous barrage of past and present unrelated information. Investigations of inhibitory cerebral functions should contribute to increasing our understanding of the normal and abnormal behavior of man.

EXPERIMENTAL BASES FOR LONG-TERM STIMULATION OF THE BRAIN IN MAN

Although ESB in man is now a recognized technique for diagnosis and therapy of specific neurological disturbances and is employed in various major medical centers around the world (Delgado, Mark, Sweet, Ervin, Weiss, Bach-y-Rita & Hagiwara, 1968), the application of brain stimulation to man has aroused controversies and criticisms, and some have considered it an aggressive procedure which should be banned from medical practice. This situation is in contrast with the wide acceptance of surgical therapies for brain problems, such as Parkinson's disease, epilepsy pain, and diskinesias. Why is surgical brain destruction accepted while brain exploration with fine wires and ESB has been questioned? Why are wires and pacemakers implanted in the heart without hesitation while the application of similar techniques to the brain is looked upon with reservations?

The source of concern may stem from traditional tabus and attitudes according to which the brain has been considered the inviolable material basis of the mind, individual personality and freedom. Fears have also been more or less explicitly expressed that ESB could introduce the nightmare of mass control of man, overriding and overpowering individual self-determination. Unfortunately, science fiction has exploited these fantasies.

As suggested briefly in this paper and discussed more extensively elsewhere (Delgado, 1969b), ESB has clear limitations and it can only trigger preestablished mechanisms and responses, whereas it cannot send ideas or instructions to a subject. A train of electrical pulses (for example, the commonly used 100 Hz, square waves, 0.5 msec, cathodal stimuli, below 1 mA) is a monotonous signal without specific information and lacking feedbacks, whose characteristics cannot be compared with the finess and significance of sensory inputs. ESB may evoke well-organized behavior or it may influence emotional reactivity by triggering stored formulas of response, but it cannot change personal identity, which depends on a vast amount of past experience. It cannot synthetize a motor act, which depends on many servo loops and the nearly instantaneous processing of sensory information, nor can it select one from a variety of possible efferent messages, or coordinate the multiplicity of effects necessary for a well-organized performance.

What are the experimental bases for using ESB as a therapeutic procedure in patients? From the medical point of view, brain stimulation should be evaluated according to the same standards as other clinical procedures, considering biological tolerance, risks involved, effectiveness, and reliability. In these respects, extensive studies in higher animals including rhesus monkeys, gibbons, and chimpanzees, as well as more limited experience in man, support the following conclusions.

1. Surgery for the implantation of electrodes involves a low risk. An electrode assembly of seven leads is approximately 1 mm in diameter, and local hemorrhage and neuronal destruction along the implantation tract are assymptomatic. The trauma of the procedure is comparable to that of a ventriculography.

2. Biological tolerance of the brain is excellent for probes of platinum, gold, and stainless steel, and also for Teflon and special insulating varnishes. Electrodes constructed of these substances may remain implanted indefinitely. Other materials, however, such as silver and copper, may cause tissue reaction and are not suitable for intracerebral implantation.

3. After a few weeks, the implanted probes are covered by a thin (0.1–0.2 mm) capsule of glia which does not impede the passage of electrical and chemical information to the probes. Immediately beyond the glia capsule, the neurons have a normal histological aspect.

4. Functional tolerance of the implanted leads is demonstrated by the absence of signs of irritation such as spikes or slow waves, and by the constancy of recorded spontaneous electrical activity.

5. Reliability of electrically evoked effects has been demonstrated during years of implantation, without changes in thresholds or impedance. Motor and behavioral responses remain constant throughout the entire experimentation time.

6. Repeated application of electrical stimulations proved safe, as evidenced by the lack of histological damage. In brain sections stained with the Klüver method, neurons of the areas stimulated many thousands of times through months of experimentation were similar in aspect to neurons of nonstimulated regions.

7. Long-term, programmed stimulation of some cerebral structures could be maintained indefinitely with persistence of the evoked effects. For example, in one monkey, electrical stimulation of the rhinal fissure four times per minute evoked a smile-like response about 500,000 times (Delgado, Rivera, & Mir, 1971).

STIMOCEIVERS IN MAN

Application by our group of stimoceiver technology in patients with psychomotor epilepsy who had electrodes implanted in the amygdala and hippocampus for therapeutic reasons has been reported (Delgado et al., 1968). The advantages of this methodology are:

1. The patient may be instrumented for telestimulation and recording simply by plugging the stimoceiver into the electrode socket on the head.

2. Use of the instrumentation does not limit or modify spontaneous behavior.

3. Without experiencing any disturbance or discomfort, the patient is continuously available day and night for intracerebral recordings or treatment.

4. Studies can be performed without introducing factors of anxiety or stress within the relatively normal environment of a hospital ward and during spontaneous social interactions among patients.

5. Cerebral explorations can be conducted in severely disturbed patients who would not tolerate the confinement of the recording room.

6. The absence of long connecting wires eliminates the risk that during unpredictable behavior or during convulsive episodes, the patient may dislodge or even pull out the implanted electrodes.

7. Programmed stimulation of the brain may be continued for as long as required for patient therapy.

Studies carried out in four patients with the stimoceiver technology support the following conclusions.

1. Depth recordings reveal local activity rather than diffuse volume conductor fields, giving anatomical accuracy to the obtained data.

2. Abnormalities in spontaneous behavior including aimless walking, speech inhibition, and psychological excitement coincided with abnormal EEG patterns.

3. In one patient, arrest reaction accompanied by after-discharge was elicited by stimulation of the hippocampus, and during 2 min, the recorded abnormalities in brain waves coincided with sensations of fainting, fright, and floating around.

4. Assaultive behavior similar to that observed during spontaneous crises was elicited in one patient by radio stimulation of the amygdala, a fact which was important in orienting therapeutic surgery.

TRANSDERMAL STIMULATION OF THE BRAIN IN MAN

As described in Section A.6, transdermal brain stimulators have been developed in our laboratory. During the past 5 years, these instruments have been tested in primates, and recently, we have reported the implantation of a transdermal unit for therapy (Delgado, Obrador & Martin-Rodriguez, 1973) in patient F.F., a 30-year-old white male, who had suffered a car accident damaging his left bracchial plexus. There was resulting paralysis of his left arm, and a phantom limb appeared, causing intolerable and incapacitating pain, unalleviated by drugs or physical therapy. When all other treatments failed, destructive brain surgery was contemplated and then, as an alternative, repeated electrical stimulation of the forebrain was suggested as a more conservative therapy. After several months of thorough clinical and psychological testing, an operation was performed under local anesthesia to implant a subcutaneous stimulator on top of the patient's head, placing its electrodes stereotaxically and bilaterally in the septum and head of the caudate nucleus. Surgical recovery was uneventful without any modification of the pain or clinical situation of the patient. Two weeks later, stimulation studies began, and 3 to 5 times a week during a tape-recorded, 1-hour session, the patient was interviewed and given psychological tests, while from an adjoining room, radio stimulations were sent intermittently to the different points in his brain. During each session, about ten stimulations were applied of about 5 sec duration, spaced 2–3 min apart, and in addition, there were two periods of 1 min each during which intermittent stimulations were sent, of 5 sec on, 5 sec off.

On random dates and for periods of up to 1 week, unknown to the patient or interviewers, no stimulations were sent during these sessions. Following several months of study, data analysis indicated that stimulation of one point evoked lasting alleviation of pain and concurrent improvement of the patient's formerly hostile behavior. At present, three years after these therapeutic stimulations, the patient remains pain-free and is gainfully employed. During this time, the subcutaneous instrumentation was comfortable for the patient, well tolerated, and reliable.

TECHNIQUES FOR CHEMICAL STUDIES

Electrical and chemical phenomena are closely related, and both must be taken into consideration for the understanding of brain physiology. In general, chemical investigation of the brain requires killing of the experimental animal and the removal of the brain in order to analyze its chemical composition. With this method, repetition of experiments in the same subject obviously is not possible, the significance of data must be evaluated statistically in different animals, and direct controls are sometimes difficult to establish. In addition, homogenates of large areas of the brain mask the neurochemical differences localized in discrete structures.

To avoid these problems, several authors have used "push-pull" cannulas for the perfusion of liquids through the brain tissue (Gaddum, 1961; see details and bibliography in Delgado, 1966; Delgado, DeFeudis, Roth, Ryugo, & Mitruka, 1972). Direct administration of drugs into the ventricles or into the brain has been used in man for therapeutic purposes (Nashold, 1959, Sherwood, 1955). The diffusion of anesthetics through a silicone-rubber membrane placed at the tip of a push-pull cannula also has been proposed for the administration of chemicals to the brain (Folkman, Mark, Ervin, Suematsu, Bach-y-Rita, & Hagiwara, 1968; Mark, Folman, Ervin, & Sweet, 1969).

Over ten years ago, we described the "chemitrode" system (Delgado, Simhadri, & Apelbaum, 1962), consisting of two small cannulas plus an array of contacts placed alongside, that permit (a) electrical recording; (b) electrical stimulation; (c) injection; (d) collection; (e) perfusion of fluids into discrete areas of the brain in unanesthetized animals. Simultaneous electrical, chemical, and behavioral studies in the conscious animal are thus shown in the review by Delgado et al., 1972).

The above methods for intracerebral injection and perfusion have several common disadvantages; namely, risk of infection, possible blocking of the cannula tips by clotting of tissue fluids, and sometimes a poor recovery rate of perfusates. In order to overcome these problems, we developed the "dialytrode" system (Delgado 1971, Delgado et al., 1972). External dialytrodes are similar to chemitrodes except that the tip of a dialytrode is enclosed by a small, porous bag which forms a barrier for microorganisms and tissue cells, while permitting the passage of fluids and chemicals. The transdermal dialytrode is a totally subcutaneous device, permitting injection and collection of fluids, as well as electrical stimulation and recording. This instrument may be suitable for therapeutic application to man.

RESULTS OF CHEMICAL STUDIES

1. *In vitro* studies were performed, immersing the dialytrode in a small bath. Synthetic spinal fluid was then circulated inside the bag at a rate of 4.0 μl/min, and labeled substances, such as ^{3}H-tyrosine, were placed in the circulating fluid or in the bath. In this way it was demonstrated that there was an outward passage of about 0.5% of tyrosine per hour, which is a substantial amount, considering that the capacity of the dialytrode bag is only 3 μl and the surface of exchange membrane is only 15 mm^2. A similar rate of inward passage was also demonstrated when ^{3}H-tyrosine was placed in the bath. The rate of exchange was similar when labeled norepinephrine was used.

2. "In vivo" studies in monkeys were performed, implanting dialytrodes in the amygdala, caudate nucleus and other brain structures. The passage of chemicals from inside the dialytrode to the brain tissue was demonstrated, perfusing L-glutamate (1 M) through the cannulas at a rate of 4 μl/min. After about 10 min, a typical glutamate seizure activity appeared in the amygdala, lasting for 2 to 6 min. Other experiments demonstrated the inward passage of metabolites from the brain to the spinal fluid circulating inside the dialytrode bag. Perfusates collected from the amygdala and analyzed with the automatic amino-acid analyzer demonstrated the presence of glutamine, asparagine, serine, glutamate, glycine, and α-alanine, indicating that these substances are released from the

amygdala in detectable amounts. In other monkeys with dialytrodes placed in the caudate nucleus, glycoproteins also appeared in the perfusate with the following substances, listed in order of higher to lower concentration: fucose, mannose, glucose, galactosamine, and galactose. These compounds were identified by using standard chemicals under similar gas-chromatographic conditions to those used in analysis of the perfusates.

In addition, biosynthesis of amino acids from labeled substances has also been demonstrated. A Ringer solution of U-14-C-D-glucose was perfused through the left amygdala dialytrode at a rate of 1.2 μl/min for one hour; then, after a quick wash out, Ringer solution was perfused at the same rate for 12 hours and the perfusate analyzed in the automatic amino-acid analyzer. Results show traces of labeled aspartate and citrulline, indicating that the injected glucose had diffused to the brain where it was converted into the two labeled amino acids in sufficient amount to diffuse back to the dialytrode.

It may be expected that changes of sugars, amino sugars, amino acids and other substances in specific brain structures should have a relation to different neurophysiological and behavioral conditions such as epileptogenic activity, sleep–wakefulness, and rage–placidity. The study of these basic problems is possible with dialytrode technology. The main advantage of the transdermal dialytrodes is that they remain functional for extended periods of time, being always available without being obstrusive when not in use. The system may provide new diagnostic and therapeutic possibilities for man, based on collection of neurochemical information from the brain and on the long- or short-term adminstration of drugs to specific cerebral structures.

LOOKING AT THE FUTURE

Electronic technology provides methods for studying, influencing, and understanding the central nervous system in animals and man. The electrical potentials of the neurons may be amplified, seen, recorded and modified, and correlated with sensory, motor, and behavioral phenomena. The use of integrated circuits and thin-film techniques have increased many fold our capability to investigate and to influence the working neurons. Radio communication with the brain permits extension to the field (Figure 3) of neurophysiological investigations which, until recently, were restricted to the laboratory.

We may expect further miniaturization and increased complexity of instrumentation to stimulate and to record brain activity. In the near future microscopic computers will be developed, small enough to be implanted under the teguments and powered by transdermal sources of energy. This instrumentation will allow, through pattern recognition, controlled stimulation and programmed

Figure 3. Gibbons with implanted electrodes in the brain, instrumented with radio stimulators and transmitters of movements, on the Hall Island (Bermuda), where experiments are conducted while the animals enjoy complete freedom.

feedbacks, the establishment of artificial electronic links between selected cerebral structures, and the modification of local reactivity by influencing the set points of physiological sensors.

At parity with electronic developments, neurochemical technology is also advancing, permitting the detection of traces of substances and the analysis of metabolic pathways of labeled substances. The possibility to introduce and to withdraw chemical information from discrete areas of the brain in conscious subjects while tests are performed, provides new means to investigate the regional neurochemistry of behavior as well as to apply new forms of therapy. Chemical blocking of hyperactive cerebral structures or chemical modification of local reactivity may be a more conservative and more effective technique than its destruction by surgery, as at present is done for the therapy of a variety of cerebral illnesses from dyskinesias to epilepsy.

Looking at the desirable future direction of brain research, we should prepare ourselves for a new age in the evolution of science in which new clinical therapies will be implemented and improved methods devised to guide the education of future man. Wisdom and ethical codes will be of primary importance as man assumes direction of the evolution of his own neurological mechanisms.

ACKNOWLEDGMENTS

The experiments mentioned in this paper were performed at Yale University School of Medicine, New Haven, Connecticut, U.S.A. and are being continued at the Autonomous Medical School, Madrid, Spain.

The studies were supported in part by the following grants: USPH MH 17408; the New York Foundation; the International Psychiatric Research Foundation; Fundación March; Fundación Rodríguez Pascual; and Instituto Nacional de Previsión.

This manuscript is a modified version of the papers presented at the Manfred Sakel Institute and at the Denghauser Group.

REFERENCES

Beritoff, J. S. Neuronal Mechanisms of Higher Vertebrate Behavior. Translated from Russian and Edited by W. T. Liberson. Boston: Little Brown, 1965, 384 pp.

Delgado, J. M. R., Simhadri, P., and Apelbaum, J. Chronic implantation of chemitrodes in the monkey brain. *Proc. Int. Union Physiol. Sci.*, 1962, *2*, 1090.

Delgado, J. M. R. Telemetry and telestimulation of the brain. In Bio-Telemetry, L. Slater (Ed.), New York: Pergamon Press, 1963, pp. 231–249.

Delgado, J. M. R. Free behavior and brain stimulation. In International Review of Neurobiology (Vol. VI), C. C. Pfeiffer & J. R. Smythies (Eds.), New York: Academic Press, 1964, pp. 349–449.

Delgado, J. M. R. Intracerebral perfusion in awake monkeys. *Arch. Int. Pharmacodyn*, 1966, *161*, 442–462.

Delgado, J. M. R. Radio stimulation of the brain in primates and in man. *Anesth. Analg.*, 1969a, *48*, 529–543.

Delgado, J. M. R. Physical Control of the Mind: Toward a Psychocivilized Society. Vol. XLI, World Perspectives Series, R. N. Anshen (Ed.), New York: Harper & Row, 1969b, 280 pp.

Delgado, J. M. R. Telecommunication in brain research. In Telemetric Methods in Pharmacology, Proceedings IV Int. Congr. Pharmacol., Vol. V. Basel: Schwabe, 1970, pp. 270–278.

Delgado, J. M. R., Johnston, V. S., Wallace, J. D., & Bradley, R. J. Operant conditioning of amygdala spindling in the free chimpanzee. *Brain Research*, 1970, *22*, 347–362.

Delgado, J. M. R. Dialysis intracerebral. In Homenaje al Prof. B. Lorenzo Velazquez. Madrid Editorial Oteo, 1971, pp. 879–899.

Delgado, J. M. R. & Mir, D. Infatigability of pupillary constriction evoked by stimulation in monkeys. *Neurology*, 1966, *16*, 939–950.

Delgado, J. M. R., Mark, V., Sweet, W., Ervin, F., Weiss, G., Bach-y-Rita, G., & Hagiwara, R. Intracerebral radio stimulation and recording in completely free patients. *J. Nerv. Ment. Dis.*, 1968, *147*, 329–340.

Delgado, J. M. R., Rivera, M., & Mir, D. Repeated stimulation of amygdala in awake monkeys. *Brain Research*, 1971, *27*, 111–131.

Delgado, J. M. R., DeFeudis, F. V., Roth, R. H., Ryugo, D. K., & Mitruka, B. M. Dialytrode for long term intracerebral perfusion in awake monkeys. *Arch. Int. Pharmacodyn.*, 1972, *198*, 9–21.

Delgado, J. M. R., Obrador, S., & Martin-Rodriguez, J. G. Two-way radio communication with the brain in psychosurgical patients. In "Surgical Approaches in Psychiatry," L. V. Laitenen & K. E. Livingston (Eds.). Lancaster, Eng.: Medical & Technical Publ. Co., 1973, pp. 215–223.

Diamond, S., Balvin, R. S., & Diamond, F. R. Inhibition and Choice. A Neurobehavioral Approach to Problems of Plasticity in Behavior. New York: Harper & Row, 1963, 456 pp.

Folkman, J., Mark, V., Ervin, F., Suematsu, K., & Hagiwara, R. Intracerebral gas anesthesia by diffusion through silicone rubber. *Anesthesiology*, 1968, *29*, 419–426.

Gaddum, J. H. Push-pull cannulae. *J. Physiol.*, 1961, *155*, 1P-2P.

Grastyan, E., Karmos, G., Vereczkey, L., Martin, J., & Kellenyi, L. Hypothalamic motivational processes as reflected by their hippocampal electrical correlates. *Science*, 1965, *149*, 91.

Hubel, D. H. & Wiesel, T. N. Receptive fields, binocular interaction and functional architecture in the cat's visual cortex. *J. Physiol.*, 1962, *160*, 106.

Jasper, H. H. & Smirnov, G. D. (Eds.) The Moscow Colloquium on Electroencephalography of Higher Nervous Activity. Suppl. 13, *EEG Clin. Neurophysiol,* 1960.

Mark, V., Folman, J., Ervin, F., & Sweet, W. Focal brain suppression by means of a silicone rubber chemode. *J. Neurosurg.*, 1969, *30*, 195–199.

Nashold, B. S. Cholinergic stimulation of globus pallidus in man. *Proc. Soc. Exp. Biol., N. Y.*, 1959, *101*, 68–80.

Quarton, G. C., Melnechuk, T., & Schmitt, F. O. (Eds.) The Neurosciences. New York: Rockefeller University Press, 1967, 962 pp.

Pavlov, I. P. Experimental Psychology. New York: Philosophical Library, 1957, 653 pp.

Sheer, D. E. (Ed.) Electrical Stimulation of the Brain. Austin: University of Texas Press, 1961, 641 pp.

Sherwood, S. L. The response of psychotic patients to intraventricular injections. *Proc. Roy. Soc. Med.*, 1955, *48*, 855–864.

Drug Effects on Foot-Shock-Induced Agitation in Mice

SAMUEL IRWIN, ROBERTA G. KINOHI, AND ELAINE M. CARLSON

Foot-shock-induced fighting was first described for rats by O'Kelly and Steckle (1939) and later used for drug studies with mice by Tedeschi, Tedeschi, Mucha, Cook, Mattis, and Fellows (1959). Our interest in this procedure arose from the observation by Valzelli (1967) that foot-shock-induced fighting was more sensitive to drug effects and easier to carry out than other procedures used for this purpose (e.g., isolation or drug-induced fighting). Also, the procedure offered means for measuring drug effects on other distress components of behavior exhibited by animals in response to foot-shock stimulation (such as vocalization, running, and leaping).

   The initial purpose of our studies was to investigate the nature and intensity of fighting and other behaviors elicited by increasing intensities of foot-shock stimulation and of drug effects on them. Later we realized that the running and leaping responses could serve as a model for the study of drug effects on "agitated" behavior.

METHOD

   The apparatus for delivering the foot shock stimulation consisted of a constant current grid ($\frac{3}{16}$" stainless steel rods; 4 per inch) adjustable from 0–1.0 mA

SAMUEL IRWIN, ROBERTA G. KINOHI, and ELAINE M. CARLSON • Department of Psychiatry, University of Oregon Medical School, Portland, Oregon.

with ±5% variation over 10^6 Ω animal resistance. Shock duration was 20 msec with 4–5 pulses per sec and a maximum grid voltage of 1500 V. For testing, the animals were placed in an open-ended transparent glass cylinder (6 × 7.75″) suspended about $\frac{1}{4}$″ above the grid as described by Tedeschi et al. (1959).

The animals used were primarily Swiss Webster female mice 6 weeks of age, tested 5 or more days after arrival. A 20% glucose solution was substituted for the food, minimally 16 hours before testing. All animals and treatments were randomized and drug studies carried out on a blind basis. Statistical analysis was with the Wilcoxon Rank Sum Test (one-tailed or two-tailed as appropriate).

METHODOLOGIC STUDIES

The effects of different intensities of foot shock stimulation (0.02, 0.04, 0.08, and 0.16 mA) were investigated after treatment with chlordiazepoxide, ethyl alcohol, chlorpromazine, methadone, imipramine, and methamphetamine. These studies revealed that a 0.08 mA intensity of foot-shock stimulation was optimal for revealing drug effects on evoked leaping, running, and fighting responses. At the 0.08 mA intensity, fighting was the most susceptible to drug-induced change followed by leaping, running, and vocalizing in that order. The intensity of vocalization was particularly resistant to change (Irwin, 1971).

The effects of saline and chlordiazepoxide (3 and 10 mg/kg orally) were investigated in both male and female Balb/C, Swiss Webster, and CFW mouse strains. Statistically significant responsiveness to the effects of chlordiazepoxide was found greatest with the Swiss Webster strain. In another study, the effects of chlordiazepoxide on 6- and 10-week-old animals of both sexes were investigated. These studies showed females at 6 weeks of age slightly more responsive to the effects of chlordiazepoxide. Accordingly, our studies were undertaken with female Swiss Webster mice at 6 weeks of age and 0.08 mA foot shock stimulation.

METHODOLOGIC PROCEDURE

On test days, animals were administered treatment and assessed with the procedure at the time of peak drug action. To assess the overall intensity of behavioral and possible adverse effects of the drugs the animals were examined just prior to testing for the degree of behavioral arousal, body tone, motor incoordination, pupil size, and palpebral closure present, rated on a 0–8 scale as described by Irwin (1968)[1]. Each animal then was placed on the grid within the class cylin-

---

[1]These data are not shown in the body of this paper but can be made available on request.

der for a 3-min measure of the duration in seconds of spontaneous locomotor activity (SMA). This was immediately followed by a 3-min period of foot-shock stimulation during which the cumulative duration in seconds of excited leaping and running (agitated) behavior was recorded. On cessation of the foot-shock stimulus, the animals assumed a freeze-like stance of postural arrest (PA) before movement was suddenly reinitiated. This interval of suppressed activity was recorded with a cutoff after 120 sec.

RESULTS

Effects of *p*-Chlorophenylalamine (PCPA) and Apomorphine

The dose-response effects of *p*-chlorophenylalamine (PCPA, a potent inhibitor of tryptophane hydroxylase which markedly reduces brain 5-HT levels) and apomorphine (a dopamine agonist) were investigated in order to determine a possible role of serotonergic or dopaminergic mechanisms in the behaviors measured.

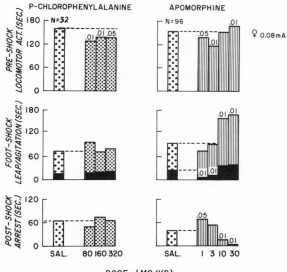

Figure 1. Comparison of the dose–response effects of *p*-chlorophenylalanine and apomorphine. The black bars denote durations of leaping in seconds; figures above the bars—statistically significant differences in response from the saline controls. Drugs were administered i.p.; testing was 24 hours after PCPA and 30 min after apomorphine.

Figure 1 shows the effects of the drugs on SMA before foot-shock stimulation, leaping and total agitated responding during the stimulation, and the duration of post-shock postural arrest (PA). As illustrated, PCPA produced a slight significant reduction of preshock locomotor activity but was without effect on any of the other measures. Apomorphine was noteworthy for producing a biphasic dose–response effect on all of the measures.[2] Doses of 1 and 3 mg/kg of apomorphine i.p. reduced the preshock locomotor activity and foot-shock-induced leaping and increased the duration of postshock PA. At these doses, apomorphine was primarily retarding and slightly anxiolytic, but with increase in dosage to 10 and 30 mg/kg, the drug became activating. It increased spontaneous activity, greatly increased the agitated responding, and virtually eliminated the period of PA following foot shock stimulation. The data, thus, ruled out serotonin but suggested the probable importance of dopamine on all of the measures studied. This observation led us to investigate the interactional effects of apomorphine with a variety of other drugs.

Interactional Effects of Apomorphine with Drugs:

Tetrabenazine and Reserpine

Figure 2 shows the results of the administration of 30 mg/kg of tetrabenazine or reserpine orally in combination with either saline or apomorphine (20 mg/kg i.p.). The study was carried out with separate groups of animals who received the drugs 2 or 20 hours prior to testing. The testing was carried out 30 min after the apomorphine injection. As may be seen, the tetrabenazine produced significant depressant effects after 2 hours with apparent return to normal levels of responding at 20 hours. Also, the administration of apomorphine completely reversed the depressant effects seen at 2 hours and produced effects almost indistinguishable from that which apomorphine alone might produce in the absence of tetrabenazine.

Reserpine alone produced near complete suppression of preshock locomotor activity and significantly prolonged post shock PA at both the 2- and 20-hour intervals of testing. Both of these effects were significantly antagonized by apomorphine, but the agitated responding to foot-shock stimulation at the 2-hour interval of testing appeared to be significantly increased rather than decreased. This may reflect a possible sustained increase in the levels of substances other than dopamine released by reserpine during this earlier period of testing, since the other measures sensitive to dopaminergic effects were depressed by reserpine. The data suggest that tetrabenazine may be more effective than reserpine in the management of agitated behavior clinically.

---

[2] These data are not shown in the body of this paper but can be made available on request.

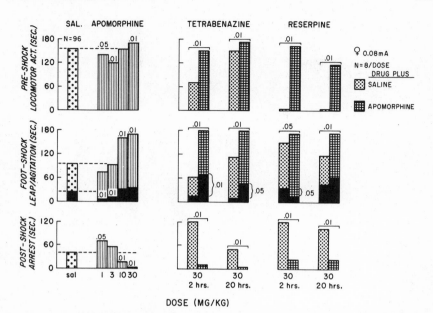

DOSE (MG/KG)

Figure 2. Interactional effects with apomorphine. Apomorphine data is as shown in Figure 1. Animals received 30 mg/kg of tetrabenazine or reserpine orally and received either saline or 20 mg/kg of apomorphine i.p. 30 min before testing (2 or 20 hours posttreatment). The figures denote the levels of statistically significant differences between the drug–saline and drug–apomorphine treatment groups.

Perphenazine, Imipramine and Chlordiazepoxide

Figure 3 shows a comparison of the effects of saline, perphenazine, imipramine, and chlordiazepoxide administered orally alone and in combination with apomorphine (20 mg/kg i.p.). As before, the animals were tested 30 min after apomorphine administration at the time of peak drug action. In this study perphenazine alone (at the 1.0 mg/kg dose only) produced significantly reduced locomotor activity and foot-shock-induced agitation, and increased postshock PA. Apomorphine, in combination, significantly antagonized the higher dose effects and significantly increased agitation and reduced the postshock PA at the 0.1 and 0.3 mg/kg doses of perphenazine. But these latter effects at the 0.3 mg/kg dose were less than those produced by apomorphine alone indicating, conversely, that perphenazine also could antagonize the effects of apomorphine. The data thus confirmed the reported ability of a phenothiazine-type major tranquilizer to block the effects of dopamine, and were congruent with the known ability of these drugs to reduce agitated behavior in man.

Imipramine alone, particularly at the 30 and 100 mg/kg doses, produced a slight but nonsignificant increase in locomotor activity and agitated responding,

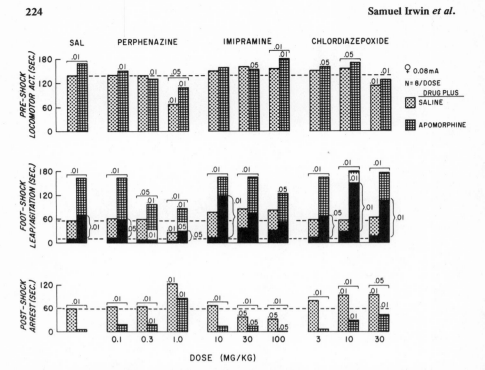

Figure 3. Interactional effects with apomorphine. Animals were tested 2 hours after per-phenazine and 1 hour after imipramine or chlordiazepoxide administered orally. Thirty minutes before testing, saline control and drug-treated animals were administered either saline or 20 mg/kg of apomorphine i.p. The figures denote statistically significant differences between the saline and apomorphine-treated drug groups.

and slightly reduced the postshock PA. It did not block or enhance the effects of administered apomorphine, which merely produced its expected effects. The data thus were congruent with the mild activating effects of imipramine observed clinically in retarded depressions, as well as with its known ability to activate psychoses, presumably through increasing responsiveness to provoking stimuli with exacerbation of agitated behavior.

Unlike our findings in similar studies, chlordiazepoxide alone had no effect on agitated responding to the foot-shock stimulation in this study. This absence of effect may have resulted from the unusually low control levels of agitated responding exhibited by the animals. However, it produced a dose-related increase in postshock PA and a slight reduction of preshock locomotor activity at the highest dose tested (30 mg/kg). Chlordiazepoxide did not block the expected augmentation of agitation following apomorphine and may even have slightly increased it, particularly in the leaping component of the response. The highest dose of chlordiazepoxide antagonized the preshock locomotor stimulant effect of apomorphine, and there appeared to be dose-related antagonism of apomorphine effects on postshock PA. The interactional effects of chlordiazepoxide with apomorphine, thus, were of a mixed type, with exacerbation of the apomorphine effects only during the foot-shock stimulation.

Comparison of Acute and Chronic Drug Effects

In these studies an effort was made to determine the altered profiles of drug action that resulted from the repeated daily oral administration of psychoactive drugs. Animals were randomly assigned to treatment groups and administered saline or drug for 13 or 20 days. On the next day following this period they received their assigned treatment, and were tested as described. Animals that had received saline received either another dose of saline or their first dose of drug, thus making it possible to compare the effects of single and repeat doses of drugs on the same test day in animals with a similar prior history of handling. In these studies there were typically 16 saline control animals and 8 animals per dose of drug tested with testing at the time of peak drug action.

Effects of Imipramine, Chlorpromazine, Perphenazine and Chlordiazepoxide

Imipramine in single doses of 30 and 100 mg/kg orally produced a slight increase in preshock SMA and foot-shock-induced agitation, and a dose-related reduction of postshock PA (Figure 4). The overall effect, thus, was one of slight activation and increased agitation in response to foot-shock stimulation. With chronic dosing, slight tolerance developed to the increase in agitation at the low dose, but there was an exacerbation of the agitation at the higher dose. With the low dose there was no significant change in either pre- or postshock behavior with chronic administration. With repeated use of the high dose, complete tolerance developed to the decrease of postshock PA seen initially.

Both the low and high doses of chlorpromazine acutely (3 and 10 mg/kg) produced a dose-related reduction of preshock SMA and foot-shock-induced agitation, and an increase in the duration of postshock PA. With chronic administration, particularly at the high dose of testing (10 mg/kg), significant tolerance developed to all the measures (behavioral, neurologic, and autonomic) except one—postshock PA. In a subsequent assessment of the same animals, we did not find any tolerance development to the ability of chlorpromazine to suppress the bizarre, stereotyped behavior produced by a 3 mg/kg dose of d-amphetamine. Our data suggest that the initial effect of chlorpromazine on agitated responding was a result of physiologic incapacity rather than from a selective effect on the agitation per se. Since the stereotyped behavior produced by d-amphetamine has been demonstrated to result from dopamine release, and since apomorphine in

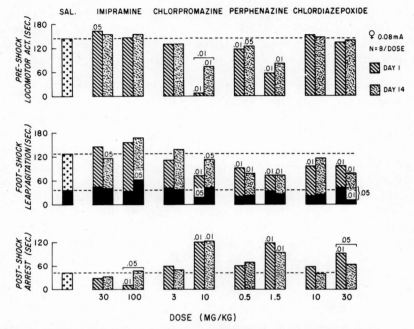

Figure 4. Comparative effects of acute and chronic drug administration. Animals were administered saline or drug over 13 days and tested after their assigned treatment on the 14th day. Treatments were administered once daily orally ; testing was 2 hours after the administration of perphenazine and 1 hour after the other drugs. Figures above the bars denote statistically significant differences from the saline controls or between the effects of acute and chronic dosing.

higher dosage could vitually abolish the postshock PA, it seems probable that tolerance does not develop to the dopamine-blocking effects of chlorpromazine with chronic dosing (as revealed by the continued suppression of amphetamine-induced steretypy and prolongation of the postshock PA. The data also suggest that the effects of chlorpromazine on these two measures (and on dopamine in particular) may be significantly related to its efficacy in the treatment of schizophrenics, since these were the only residual effects following chronic dosing.

The dose–response effects of perphenazine were found to be similar in profile to that of chlorpromazine, except that no significant tolerance developed with chronic daily dosing. Also, perphenazine was effective in reducing the foot-shock-induced agitation in doses that produced very little observable behavioral or physiologic changes. The data are congruent with what has been observed clinically in that perphenazine can reduce agitation in patients at doses that evoke minimal overt behavioral or physiologic effects. Unlike chlorpromazine, this capacity for reducing agitation is not lost with chronic dosing.

Chlordiazepoxide, at both the 10 and 30 mg/kg doses, produced a significant reduction of foot-shock-induced agitation acutely to which tolerance did not develop with chronic dosing; also, a very slight but nonsignificant reduction of SMA occurred at the 30 mg/kg dose. Like the major tranquilizers, it too produced a significant dose-related increase in postshock PA. But unlike the major tranquilizers, significant tolerance developed to this effect with chronic dosing. Parenthetically, the interactional studies with chlordiazepoxide and apomorphine revealed a mutual pharmacodynamic antagonism between their effects on postshock PA, thus suggesting that the initial increase with this measure may have resulted from an antagonism of apomorphine to which tolerance developed with chronic dosing.

Effects of Meperidine, Morphine, Methadone and Cyclazocine

A study comparing the effects of the narcotic analgesics meperidine, morphine, and methadone, and the narcotic antagonist cyclazocine is shown in Figure 5. As may be observed, the drugs possessed mixed stimulant and depressant

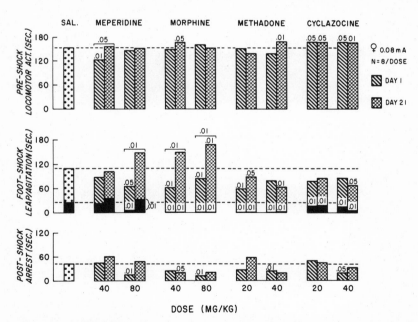

Figure 5. Comparative effects of acute and chronic administration. Animals were administered either saline or drug over 20 days and tested after their assigned treatment on the 21st day. Treatments were administered once daily orally; testing was at the time of peak drug action (30 min posttreatment). Figures above the bars denote statistically significant differences from the saline controls or between the effects of acute and chronic dosing.

effects. In doses that had very little effect on preshock locomotor activity, both meperidine and morphine significantly decreased the foot-shock-induced agitation and significantly decreased the postshock PA. With chronic administration, however, not only did complete tolerance develop to their effects on agitation, but a significant rebound increase in agitation was seen. Also, significant tolerance developed to the effects of meperidine (but not with morphine) on the foot-shock-induced leaping and postshock PA.

The pattern of effects of methadone and cyclazocine on agitated responding was different from that of meperidine and morphine, for with these agents, tolerance with continued dosing did not develop to their effects on agitation and postshock PA. Of the four agents, only cyclazocine significantly increased preshock locomotor activity acutely, an action to which tolerance did not develop. In general, with chronic dosing, morphine and cyclazocine appeared to be the most activating drugs on preshock locomotor activity; and methadone and cyclazocine the most effective for reducing agitated responding. Tolerance development with chronic dosing appeared to be greatest with meperidine.

Particularly interesting with both the narcotic analgesics and narcotic antagonists, to a degree not seen with any other psychoactive drug that we have studied, was the marked reduction of foot-shock-induced leaping seen. The effects may have been a consequence of their analgesic activity. Because initial postshock PA also was reduced, the reduction of leaping responses could not be attributed to any antagonism of dopamine. The consequence of such a block would be a prolongation of the postshock PA rather than the reduction in its duration seen with the narcotics.

Effects of Amitriptyline and Perphenazine and Their Combination

Shown in Figure 6 is a summary of the results obtained with two dose levels each of amitriptyline and perphenazine and their combination in every possible dosage permutation. Amitriptyline is a drug which, in both animal and human studies, is highly sedating—considerably more so than imipramine. Indeed, its profile of action tends to share many of the properties of the more sedative major tranquilizers. Thus, with increase in dosage, it was observed acutely to similarly reduce spontaneous locomotor activity and prolong postshock PA, while imipramine reduced the latter. But, as with imipramine, anitriptyline also significantly increased foot-shock-induced agitation; this effect similarly increased with chronic dosing (accompanied by a significant increase in leaping behavior). Also with chronic dosing, there was a further prolongation of the postshock PA. Tolerance clearly did not develop to this effect.

The effects of perphenazine were as previously described, with a dose-related decrease in SMA and a significant prolongation of postshock PA. But there was no reduction of foot-shock-induced agitation, possibly because of the unusually

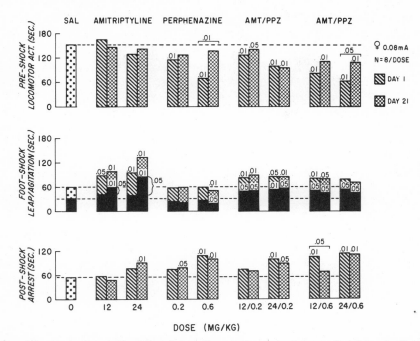

Figure 6. Acute-chronic drug-drug interaction studies. Animals were administered saline or drug treatment (amitriptyline or perphenazine, alone or in their varying dose combinations) for 20 days and tested after their assigned treatment on the 21st day. Treatments were administered once daily orally; testing was at the time of peak drug action (1 and 2 hours after amitriptyline or perphenazine respectively.) Figures above the bars denote statistically significant treatment differences from the controls and between the effects of acute and chronic dosing.

low control baseline levels of agitation present in this study. With chronic administration, however, significant tolerance did develop in this study to the effect of perphenazine on preshock SMA. As with the previous study, however, no tolerance developed to its effect on foot-shock-induced agitation or post-shock PA.

In the combination of the low dose of perphenazine (0.2 mg/kg) with the high dose of amitriptyline (24 mg/kg) only, one sees a synergism with the effects of perphenazine on pre- and postshock activity without significant interference by perphenazine with the agitation-increasing effect of amitriptyline. With chronic administration, however, the 0.2 mg/kg dose of perphenazine blocked the further increase in agitation produced by the higher dose of amitriptyline. This curtailing of agitation occurred without further affecting the other dimensions of measurement.

With the high dose of perphenazine (0.6 mg/kg) in combination with the low and high dose of amitriptyline, no synergism of the effects of perphenazine on the pre- and postshock activity was seen, but there was a clear reduction of the increased agitation produced by amitriptyline alone (both initially and after chronic administration). With repeated dosing of this combination, significant tolerance developed to the initial depression of preshock SMA similar to that seen with perphenazine alone.

The ratios of dose combinations used in this study were identical to this combination marketed as "Etrafon" or "Triavil." The data suggest that amitriptyline in combination with low doses of perphenazine synergized the locomotor depressant effects of perphenazine without significant effect on the capacity of amitriptyline to enhance the agitated responding to foot-shock stimulation. In the combination of amitriptyline with the higher dose of perphenazine, the psychomotor retarding effects of perphenazine on pre- and postshock responding remained unaltered while the activating effect of amitriptyline on agitation was still present but partly reduced. The combination, thus, would appear to be a rational one for the purposes intended, with each drug seemingly reinforcing or only partially reducing some of the desired effects of the other.

Effects of Meprobamate, Benactyzine, and Their Combination

Figure 7 shows a summary of the results with two dose levels each of meprobamate (75 and 150 mg/kg) and benactyzine (3 and 10 mg/kg) and their combination in every possible dosage permutation. The data with meprobamate and benactyzine show no significant effects acutely or chronically on preshock locomotor activity: a very small, nonsignificant reduction of foot-shock-induced agitation at the higher doses and a statistically significant increase in postshock PA with meprobamate only. On combination, however, the profile of effects became different than that with either drug given singly. Combination of the high dose of meprobamate with the low dose of benactyzine resulted in a significant reduction of preshock SMA, a slight but nonsignificant reduction of foot-shock-induced agitation, and a significant prolongation of postshock PA with little or no tolerance development to these effects with chronic dosing. Similar effects were also observed upon combination of the low or high dose of meprobamate with the high dose of benactyzine. However, as may be noted, a significant increase in foot-shock-induced agitation was seen initially upon combination of the low dose of meprobamate with either the low or high dose of benactyzine. Tolerance developed to this effect with chronic dosing. These results were single doses seemed to parallel observations we made on normal humans with similar dose combinations. In both studies, a unique profile of action emerged from combination of the two drugs.

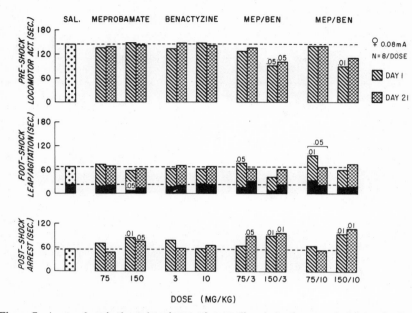

Figure 7. Acute–chronic drug–drug interaction studies. Animals were administered saline or drug treatment (meprobamate or benactyzine alone or in their varying dose combinations) for 20 days and tested after their assigned treatment on the 21st day. Treatments were administered once daily orally; testing was 60 min posttreatment. Figures above the bars denote statistically significant treatment differences from the controls and between the effects of acute and chronic dosing.

Pattern of Drug Effects

Table I shows the pattern of drug effects obtained from a broad range of dosage and further studies not discussed in this paper. Differences in the response observed between the low and high doses investigated are separated by a diagonal. The letter M denotes mixed increased (+) or decreased (−) effects; and the letter T, tolerance development with chronic administration. The plus (+) or minus (−) signs denote the direction of effects produced by the drugs, i.e., the relative increase or decrease.

The drugs under each class investigated were as follows: psychomotor stimulants (cocaine, caffeine, *d*-amphetamine, methamphetamine, methylphenidate, and magnesium pemoline); tricyclic antidepressants (nortriptyline, amitriptyline, desipramine, and imipramine); catecholamine-releasing major tranquilizers (tetrabenazine and reserpine); dopamine-blocking major tranquilizers (promazine, chlorpromazine, thoridazine, chlorprothixine, acetophenazine, thiothixene, tri-

Table I. Pattern of Drug Effects[a]

Drug type	Leaping		Agitation		Postural arrest	
	Acute	Chronic	Acute	Chronic	Acute	Chronic
Apomorphine	–/+		–/++		+/=	
Psychomotor stimulants	M/+		+		+/–	
Tricyclic antidepressants	±/M	+	+	++	+/M	T[b]
C.A. release major tranquilizers	+		±		++	
Dopamine-block major tranquilizers	±/∓	0	–	–	++	++
Minor tranquilizers	–	–	–	–	+	0
Sedative–hypnotics	–		0		0	
Narcotic analgesics	=	= or T	–	– to +	–	– or T
Narcotic antagonists	=	=	0/–	–	– to +	0
α-Adrenergic blockers	M		0		±	
Anticholinergics	0		0		/±	
Antihistamines	+/–		0		/–	
p-Chlorophenylalanine	0		0		0	

[a] low/high dose; M = mixed – or + response; T = tolerance development. ±, + and ++ or ∓, – and = represent very slight, moderate, and marked increases or decreases in response.
[b] Only when lowered PA scores present initially.

fluoperazine, fluphenazine, and haloperidol); minor tranquilizers (diazepam, chlordiazepoxide, oxazepam, meprobamate, and ethyl alcohol); sedative–hypnotics (phenobarbital and pentobarbital); narcotic analgesics (morphine, methadone, and meperidine); narcotic antagonists (pentazocine, nalorphine, and cyclazocine); α-adrenergic blockers (phentolamine and phenoxybenzamine); anticholinergics (atropine and scopolamine); and antihistaminics (chlorpheniramine and pyribenzamine). The effects of ethyl alcohol most resembled those of the minor tranquilizers, thus determining our classification.

As previously noted in the text and as shown in Table I, apomorphine in lower dosage produced a slight reduction of leaping and agitated responding to foot-shock stimulation and an increase in postshock PA. In high doses it produced a profile of effects similar to that observed with the psychomotor stimulants and tricyclic antidepressants, characterized by increased leaping and agitated responding to foot-shock stimulation and a reduction of the postshock PA interval. The catecholamine-releasing major tranquilizers (reserpine and tetrabenazine) produced similar effects except that increasing doses greatly prolonged rather than decreased the postshock PA. Of interest too was that all of the agents, including the psychomotor stimulants, produced a small increase in postshock PA in the lower doses tested. Of all the psychomotor stimulants studied, however, magnesium pemoline was virtually without effect on the various measures over a dosage range of 3–30 mg/kg. Its profile of action, thus, differed from that of the other psychomotor stimulants.

The dopamine-blocking major tranquilizers (phenothiazines, thioxanthines, and butyrophenones) had little effect on leaping behavior, but primarily reduced the vigor and duration of agitated running in response to foot-shock stimulation and considerably increased the postshock PA (to which tolerance did not develop with chronic administration). Of the various agents investigated, promazine was the least effective in either reducing the agitated responding or prolonging the postshock PA, requiring doses of 30 mg/kg for effect. Thiothixine was the next least effective. Unlike the other major tranquilizers, where a close correspondence exists between animal and human dosage for behavioral effects, we found it necessary to give doses as high as 10 mg/kg to obtain observable effects with thiothixine in mice, rats, cats, and monkeys. The human is apparently much more dose-responsive to this agent. It was observed too that almost complete tolerance developed to the initial effects of chlorpromazine on foot-shock-induced agitation but not with perphenazine, suggesting that the effect of chlorpromazine on agitated responding was nonspecific and more likely a consequence of the profound weakness and stupor produced by it at the doses required for effect.

The minor tranquilizers and ethyl alcohol were found to significantly reduce both the leaping and overall agitated responding to foot-shock stimulation and to prolong the postshock PA interval. With the chronic administration of chlordiazepoxide, also, tolerance was observed to develop only to its effects on postshock PA. Thus, the minor tranquilizers appeared to be effective on both components of agitated responding (leaping and running), while the major tranquilizers were only effective in reducing the vigor and duration of agitated running responses. The magnitude of effects of the major tranquilizers on postshock PA, however, was greater than with the minor tranquilizers and tolerance to this effect did not develop.

The sedative–hypnotics were found to reduce only the leaping response to foot-shock stimulation, and to be without effect on the other components of responding. Thus, they were not as effective as the minor tranquilizers on the agitated responding.

The narcotic analgesics and narcotic antagonists were uniquely different in profile. Although similar in effect to the minor tranquilizers on leaping and agitated responding, they produced a much greater reduction of the leaping responses and decreased rather than increased postshock PA. Tolerance to the effects on leaping developed with meperidine but not with morphine, methadone, or cyclazocine. With both meperidine and morphine, a rebound increase in foot-shock-induced agitation was seen after chronic dosing; with methadone or cyclazocine, a continued decrease was observed.

As seen at the bottom of Table I, minor or only negligible effects were obtained with α-adrenergic blocking agents, anticholinergics, antihistaminics, and p-chlorophenylalanine, a depletor of 5-HT.

DISCUSSION

Our studies showed an increase in foot-shock-induced agitation (leaping and running) and postshock PA with increasing intensities of stimulation. Also, there was a casual rather than specific inverse linear relationship between the levels of preshock locomotor activity and postshock inactivity. They showed that whenever spontaneous locomotor activity increased, one was apt to see a decrease in the interval of postural arrest (PA) following foot-shock stimulation and vice versa.

Our data suggested that the postshock PA might be primarily under dopaminergic control since apomorphine, a dopamine agonist, could virtually reduce the postural arrest to zero while dopamine antagonists such as perphenazine or chlorpromazine greatly prolonged it. Tolerance to this effect did not develop with chronic dosing.

It was evident from our data that apomorphine also could greatly increase foot-shock-induced agitation (leaping and running), but that the major tranquilizers in reasonable dosage could only partially abolish or ameliorate these modes of response (with or without apomorphine). Accordingly, it seemed apparent that substances other than dopamine might play a role in the agitated responding (possibly other biogenic amines).

A number of other agents such as the psychomotor stimulants and tricyclic antidepressants increased foot-shock-induced agitation. The agitated responding was seen to further increase with chronic administration of the tricyclic antidepressants and to increase also after the chronic administration of meperidine and morphine. Both classes of drugs are of proven clinical effectiveness in the treatment of retarded depression. Whether or not the psychomotor stimulants or apomorphine retain their effects on agitation with chronic dosing is at present unknown, but the data suggest that the capacity to increase and sustain agitated responding to foot-shock stimulation with chronic administration may reflect drug effects on neurotransmitter mechanisms of value for the treatment of retarded depressions.

Finally, it was found that only narcotic analgesics or narcotic antagonists with agonistic properties could markedly reduce or completely abolish the leaping (vertical) component of agitated foot-shock responding, and that this effect was independent of the direction of drug effects on postshock PA (as may be seen in Figure 5 following the chronic administration of 20 mg/kg of methadone).

What this procedure provides for study, therefore, is a model for investigating

Table II. Therapeutic Correlates of Drug Action[a]

Therapeutic action	Leaping	Agitation	Postural arrest	Drug type
Anti-agitation	0	–	++	Major tranquilizer
Anxiolytic	–	–	0	Minor tranquilizer
Anaglesic/anxiolytic	=	–	0 to –	Methadone
Analgesic/activating	=	+	–	Morphine
Agitation/activating	+	++	– or ±	Tricyclic antidepressants

[a]0 denotes no effect; – or + symbols denote the direction and magnitude of effects produced.

drug effects on agitated behavior in mice that seems highly correlated with clinical experience with these agents. Also, it offers a means for investigating neurotransmitter mechanisms underlying both the agitated response and postural arrest following foot-shock stimulation.

We believe that drug effects on the agitated responding along with their effects on postshock PA can serve as a useful predictor of drug effects in man. As shown in Table II, the abolition of leaping responses to foot shock stimulation can be a predictor of potent analgesic activity—marked prolongation of postural arrest following foot-shock stimulation, a good predictor of psychomotor retardation and possible "antischizophrenic" activity (through the mechanism of dopamine blockade); decreased agitation with marked postural arrest as with the major tranquilizers, potent antiagitation activity; and, when associated with little or no postural arrest, as with the minor tranquilizers, more direct anxiolytic activity. Conversely, increased agitation without tolerance development on chronic dosing may be predictive both of an exacerbation of agitation and possible usefulness in the treatment of retarded depression.

In drug studies for prediction in man, we believe that it is necessary to compare the acute and chronic effects of drugs to determine whether the effects seen initially continue to persist with continued dosing. Our studies showed that tolerance does not develop to the effects of the major tranquilizers on the postshock PA or, with the more potent major tranquilizers, on the foot-shock-induced agitation. Also, the increase in agitation seen with the tricyclic antidepressants (imipramine and amitriptyline) further increased with chronic administration, possibly as tolerance developed to some of their more depressant effects.

SUMMARY

Drug effects were investigated on the agitated (leaping and running) response of animals to 0.08 mA foot shock stimulation over 3 min, and on the duration of a postural arrest phenomenon (PA) seen immediately after cessation of the foot-shock stimulus. Many drugs could be classified by their pattern of effects on these measures. A high correlation was found between the ability of drugs to increase or decrease agitated responding in this procedure and the reported effects of drugs on anxious or agitated behavior in patients. A fair correlation also was found between the prolongation or reduction in the duration of PA and the production of psychomotor retardation or activation by drugs in man. For example, major tranquilizers characteristically prolonged the PA and reduced the running but not leaping component of the agitated responses to foot-shock stimulation. The minor tranquilizers reduced both components of agitated responding and, after chronic dosing, had no effect on the postshock PA. The tricyclic antidepressants increased agitated responding without constant or predictable effects on postshock PA. The narcotic analgesics reduced both components of agitated responding, most prominently the leaping response, and also tended to reduce postshock PA.

The postshock PA response seemed to be largely under dopaminergic control. Apomorphine, a dopamine agonist, could virtually abolish the response, while major tranquilizers (dopamine-blocking agents) effectively blocked the effects of apomorphine and greatly prolonged the postshock PA when given alone. Our data showed that while apomorphine could also greatly increase both components of agitated responding to foot-shock stimulation, the effect of drugs on the agitated responding was apparently influenced by nondopaminergic mechanisms as well.

The procedure is recommended as a useful model for predicting drug effects on both anxious or agitated behavior, and their possible psychomotor retarding or activating effects in man, with reasonable inference as to whether the drug would prove to be effective in the treatment of schizophrenic reactions or retarded depressions.

ACKNOWLEDGMENTS

We wish to thank the following pharmaceutical companies for contributing the drugs for these studies: Abbott Laboratories (pentobarbital), Burroughs Wellcome and Co. (methamphetamine), Geigy Pharmaceuticals (imipramine), Hoffman-La Roche Inc. (chlordiazepoxide), The Lilly Research Laboratories (methadone), and Schering Corporation (perphenazine). This study was supported by USPHS Research Grant MH 10990, National Institute of Mental Health.

REFERENCES

Irwin, S. Comprehensive observational assessment: Ia. A systematic, quantitative procedure for assessing the behavioral and physiologic state of the mouse. *Psychopharmacologia (Berl.)*, 1968, *13*, 222–257.

Irwin, S. Drug effects on distress-evoked behavior in mice: Methodology and drug class comparisons. *Psychopharmacologia (Berl.)*, 1971, *20*, 172–185.

O'Kelly, L. W., & Steckle, L. C. A note on long enduring emotional responses in the rat. *J. Phychol*, 1939, *8*, 125–131.

Tedeschi, R. E., Tedeschi, D. H., Mucha, A., Cook, L., Mattis, P. A., & Fellows, E. J. Effects of various centrally acting drugs on fighting behavior of mice. *J. Pharmacol. Exp. Ther.*, 1959, *125*, 28–33.

Valzelli, L. Drugs and aggressiveness. In Advances in Pharmacology *5*, New York: Academic Press, pp. 79–108.

Indole Hallucinogens as Animal Models of Schizophrenia

EDWARD F. DOMINO

To say the least, the literature on indole hallucinogens and schizophrenia is confusing. This is because investigators who have access to schizophrenic patients do not have sophisticated gas chromatographic and mass spectrographic facilities, and those chemically sophisticated researchers who have such analytic facilities do not have access to a suitable patient research population. This situation has existed for some time and probably will continue to exist for a long time. The current nature of psychiatric research strategy, funding, social and legal pressures regarding informed consent and commitment of the mentally ill all combine to retard progress. This situation has become so confusing in the State of Michigan in the last two years that most research with mental patients has stopped. In attempting to protect the rights of the mentally ill, their rights to proper diagnoses and better treatments are being discarded.

The interested reader can find a large literature in which various fluorescent, paper, thin layer and/or gas-chromatographic techniques have been utilized to determine the presence of hallucinogenic indoles such as tryptamine (T), dimethyltryptamine (DMT) and 5-hydroxydimethyltryptamine (bufotenin, BFT) in various body fluids. Almost all investigators agree that very small amounts of such indoles, if any, are present in human fluids. Inasmuch as most chemical methods have borderline sensitivity in detecting microgram amounts of such

EDWARD F. DOMINO • Department of Pharmacology, Lafayette Clinic, Detroit, Michigan; and University of Michigan, Ann Arbor, Michigan.

compounds, it is little wonder that the subject is confusing. Emphasis should be placed on the reliability of the methods used in any particular study. Much of the older literature is on the presence of hallucinogenic indoles in the blood and urine of schizophrenic patients with and without administration of L-methionine, L-tryptophan (TP) and/or MAO inhibitors using nonspecific chemical assays. Rather than go over what has already been thoroughly reviewed elsewhere, this presentation will concentrate on five critical questions and offer evidence, pro or con, that bears on them.

CAN KNOWN HALLUCINOGENIC INDOLES BE MADE BIOSYNTHETICALLY IN LIVING ORGANISMS, ESPECIALLY IN MAN?

In Figure 1 are summarized some of the biosynthetic pathways proposed in the formation of known hallucinogenic indoles, but not pyridoindoles. The ultimate source of indoles is the essential amino acid TP. An individual's diet is the major source of TP, although it can be made biosynthetically via chorismate and several intermediates to indole-3-glycerol-phosphate and thence to TP. The metabolism of TP is complex. Most of the TP is utilized by cells to form essential proteins. About 1% of the TP is hydroxylated to form 5-hydroxytryptophan (5-HTP). Another portion is converted in the liver by tryptophan oxygenase (L-tryptophan: oxygen oxidoreductase, tryptophan pyrrolase) to form kynurenine (KYN), quinolate, and thence nicotinate, the essential vitamin B_3 necessary to form the cofactors NAD and NADP. A very small portion of TP is decarboxylated by aromatic amino acid decarboxylase (AAAD) to T. The same enzyme also decarboxylates 5-HTP to 5-hydroxytryptamine (5-HT, serotonin).

T, BFT, and DMT are hallucinogenic in man. These substances are constituents of various snuffs taken by certain South American Indian tribes. The evidence, pro and con, that some of these compounds are hallucinogenic, as well as proof that they are constituents of various snuffs, has been elegantly summarized by Holmstedt and Lindgren (1967). These substances have significant autonomic and motor effects. There is some question as to whether BFT is hallucinogenic (Isbell, 1967). Studies in animals indicate that 5-methoxy DMT (5-MODMT) is more behaviorally toxic than DMT. Hence, 5-MODMT is of special significance. Shulgin has been quoted as saying that 5-MODMT is a potent hallucinogen in man but to my knowledge nothing has been published on this.

It is a paradox that nature should provide us with the biosynthetic pathways for making hallucinogenic indoles. Axelrod (1962) first reported the presence of a nonspecific N-methyltransferase (NMT) in rabbit lung capable of forming

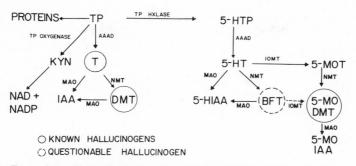

Figure 1. Formation of hallucinogenic amines from tryptophan metabolism. TP an essential amino acid is derived primarily from the diet. Most of it is converted to proteins. Via TP oxygenase, which is mainly in the liver, it is biotransformed through a series of steps to KYN and ultimately to the essential cofactors NAD and NADP. About 1% of TP is converted via TP hydroxylase (TP HXLASE) to 5-HTP. AAAD converts both TP and 5-HTP to T and especially 5-HT. Normally, type A and B MAO biotransform T to the corresponding aldehyde. Aldehyde dehydrogenase converts it to indoleacetic acid (IAA). Type A MAO biotransforms 5-HT and in a similar manner 5-hydroxyindoleacetic acid (5HIAA) is formed. NMT also can convert T to DMT and 5-HT to BFT. These tertiary amines are poorer substrates for MAO than their corresponding primary amines. Nevertheless they are converted to IAA and 5-HIAA. Indole-O-methyltransferase (IOMT) can also convert 5-HT to 5-methoxytryptamine (5-MOT). MAO can then oxidize 5-MOT and ultimately 5-methoxyindoleacetic acid is formed. The known hallucinogens T and 5-MODMT are enclosed in solid circles and BFT, a questionable hallucinogen in an interrupted circle.

N-methyltryptamine (NMET) from T and S-adenosylmethionine (SAM), and DMT from NMET and SAM. If ^{14}C-SAM is used, one can easily measure labeled indole products. However, other substances can also be methylated, so great care must be taken to prove the labeled product is indeed ^{14}C-DMT. Both Mandel (1975) and ourselves have shown, by TLC separations, that T is converted primarily to NMET. When NMET is used, it is readily converted to DMT by the rabbit-lung enzyme. Amazingly, Axelrod's important observation was not pursued for about 7 years. Then other investigators observed a similar enzyme as well as a more specific indole amine-N-methyltransferase (INMT) in many tissues of both animals and man.

In addition to SAM, 5-methyltetrahydrofolic acid (MTHF) has been reported to be an alternative methyl donor for forming DMT (Banarjee & Snyder, 1973; Hsu & Mandell, 1973). However, Mandel, Rosegay, Walker, VandenHeuvel, and Rokach (1974) have been unable to show that it acts as a methyl donor to form DMT. Instead, MTHF methylates indoles to form β-carbolines. Barchas, Elliot, DoAmaral, Erdelyi, O'Connor, Bowden, Brodie, Berger, Renson, and Wyatt (1974) have reported similar findings.

Mandel, Rosenzweig, and Kuehl (1971) reported that INMT from rabbit lung had a K_m for T of 3.3×10^{-4} M and for NMET of 5×10^{-5} M. Mandel,

Ahn, VandenHeuvel, and Walker (1972) reported that with the purified human-lung enzyme NMET had the lowest K_m of several indole amine substrates (2.8×10^{-4} M) and that the products of the reaction, BFT and DMT, were inhibitory. Saavedra, Coyle, and Axelrod (1973) reported that for rat-brain NMT the K_m for both T and NMET is 2.8 and 3.7×10^{-5} M and for SAM 5.1×10^{-5} M. Research in our laboratories at the Lafayette Clinic has indicated that K_m for the rabbit lung NMT for T is approximately 7.7×10^{-4} M and for NMET 1.5×10^{-4} M. This would indicate that high tissue concentrations of these substances are needed for the rabbit, at least, to make DMT *in vivo*. T given intracisternally to rats is converted to NMET as well as DMT (Saavedra and Axelrod, 1972). Furthermore, Ahn, Walker, VandenHeuvel, Rosegay, and Mandel (1973) have shown that the rabbit can make DMT from NMET *in vivo*. It is especially important to note that most researchers agree that in human tissues the activity of either the nonspecific NMT or the more specific INMT enzyme is very low. Our own studies using brain tissue obtained from autopsy of deceased chronic schizophrenic, organic brain syndrome or mentally normal patients who died of various physical causes has been very sobering. We have only been able to find extremely low levels of NMT activity in the human. Furthermore, we have been unable to find a difference in regional NMT activity among those three groups of patients (Domino, Krause, & Bowers, 1973). Narasimhachari, Plant, and Himwich (1972) reported enhanced NMT activity in the serum of most acute and some chronic schizophrenic patients. Wyatt, Saavedra, and Axelrod (1973d) have also reported enhanced NMT activity in the platelets of schizophrenic patients compared to mentally normal controls. When the enzyme from both the normal controls and the schizophrenic patients was dialyzed, it showed similar activity, suggesting that the schizophrenic platelets lacked a small dialyzable inhibitor. Demethylated SAM, better known as *S*-adenosylhomocysteine (SAH), immediately comes to mind as a possibility inasmuch as this agent is a potent NMT inhibitor as discussed below.

It is well known that the amine substrates for the methylated indoles exist in various tissues of animals and man. These include 5-HT and T. In addition, living organisms contain adequate levels of the *N*-methyl donors SAM and MTHF. After administering the MAO inhibitor isocarboxazid, dog brain and spinal cord T is increased two- to three-fold (Martin, Sloan, Christian, & Clements, 1972).

Conclusion

It seems fairly certain that the biosynthetic pathways for making hallucinogenic idoles exist in many tissues in animals and man. The nonspecific NMT enzyme has a rather high Km for substrates such as T and NMET, indicating that rather high tissue levels need to be present in order for DMT to be formed. Furthermore, in man the NMT and INMT enzymes have rather low levels of activity. Hence, although the biosynthetic pathways for the formation of indole hallucinogens exist, these pathways probably are not very active unless the patient has high levels of substrate in tissues containing these enzymes. The enhanced platelet NMT activity in schizophrenic patients is due to a decrease of a dialyzable endogenous inhibitor. The fact that SAH is a potent product inhibitor of this enzymatic reaction suggests that this endogenous substance should be measured in platelets of schizophrenic patients and compared with normals.

DO HALLUCINOGENIC INDOLES REALLY EXIST IN THE BODY FLUIDS OF SCHIZOPHRENIC PATIENTS IN SUFFICIENT CONCENTRATIONS TO BE PSYCHOTOGENIC?

Not long ago the answer to this question would have been a qualified yes (Franzen & Gross, 1965; Gross and Franzen, 1965; Fisher, 1968; Rosengarten, Szemis, Piotrowski, Romaszewska, Matsamoto, Stencka, & Jus, 1970; Narasimhachari, Heller, Spaide, Haskovec, Fujimori, Tabushi, & Himwich, 1971a; Narasimhachari, Heller, Spaide, Haskover, Meltzer, Strohlevitz, & Himwich, 1971b; Greenberg, 1973; Himwich, Narasimhachari, Heller, Spaide, Haskovec, Fujimori, & Tabushi, 1973; Juntenen, Struck, Warner, Frohman, & Gottlieb, 1974). Now the answer seems a qualified "no"—at least for the majority of schizophrenics because of the recent publications of Mandel et al. (1974) who used a GC-MS isotope dilution assay specific for DMT (Bidder, 1974; Lipinski, Mandel, Ahn, VandenHeuvel, & Walker, 1974; Walker, Ahn, Albers-Schonberg, Mandel, & VandenHeuvel, 1973; Wyatt, Mandel, Ahn, Walker, & VandenHeuvel, 1973a, b). Their studies indicate that a large percentage of acute and chronic schizophrenic patients have no significant plasma levels of DMT to the minimum detection level of their assay, which was 0.5–1.8 ng/ml of plasma. However, there was a very sobering observation made by these researchers who gave DMT in hallucinogenic doses to mentally normal volunteers (Kaplan, Mandel, Stillman, Walker, VandenHeuvel, Gillin, & Wyatt, 1974). They were only able to measure very low plasma levels of DMT for a very short time using the same GC-MS assay method, indicating that this hallucinogen disappears from the plasma very rapidly. This finding in man is in agreement with data in rats using a less specific fluorometric method (Cohen & Vogel, 1972). The fact that a small percentage of schizophrenic patients have detectable plasma levels of DMT indicates that they probably produce endogenously very large amounts of this substance. Investigators who have used chromatographic assays have consistently reported enhanced blood or urine levels of DMT and related indoles (Narasimhachari et al., 1971a,b; Greenberg, 1973; Himwich et al., 1973; Juntunen et al., 1974). It would appear that these latter investigators are measuring not only DMT but also other amines with similar retention times. Hence, the issue of whether schizophrenic patients have enhanced hallucinogen levels is still wide open. It is especially important to measure T, BFT, DMT, and 5-MODMT simultaneously before and during TP, and methionine loading with and without an MAO inhibitor in both normals and schizophrenic patients of all types before this question can be answered definitively.

Conclusion

Investigators who have used nonspecific chemical assay methods for measuring DMT and related substances report enhanced levels in the blood or urine of schizophrenic patients. However, when a very specific isotope dilution GC-MS technique for DMT was used, the vast majority of schizophrenic patients did not have enhanced plasma DMT levels. Further studies are obviously indicated not only because other amines may be involved but also because of low plasma levels and very rapid disappearance of DMT given in halluciongenic doses to mentally normal controls. Schizophrenic patients in whom DMT is detected in their blood obviously must synthesize enormous amounts endogenously.

DOES TOLERANCE OCCUR TO DMT?

Even if DMT were present in schizophrenic patients, it might be argued that tolerance would occur to this hallucinogen similar to LSD-25. This argument was used previously—that even if LSD-25 could be made biosynthetically it could not be an endogenous hallucinogen. However, the tolerance argument is not valid, for cyclic tolerance to LSD-25 has been observed, at least in the goat (Koella, Beaulieu, & Bergen, 1964). Cyclic tolerance might, in fact, explain why schizophrenic patients have remissions and exacerbations of their illness. Furthermore, schizophrenic patients could have disturbed tolerance mechanisms. Hence, the tolerance argument is not particularly critical as to whether an endogenous hallucinogen is psychotogenic. Still it is of interest to ask if DMT produces tolerance. Gillin, Cannon, Magyar, Schwartz, and Wyatt (1973) reported that tolerance does not occur to daily DMT in cats. However, they neglected a basic pharmacologic principle, emphasized by Seevers and Deneau (1963) that in order to produce tolerance to a substance one needs to maintain high blood levels of the chemical 24 hours a day for many weeks. Hence, we decided to re-evaluate this problem in our laboratory using hungry albino Holtzman rats trained to bar press for a sugar-sweetened milk reward. When DMT is given in a dose of 10 mg/kg intraperitoneally, bar pressing behavior in the rat stops for a short time in a dose related manner (Kovacic and Domino, 1973). When DMT was given every 2 hours for 21 days, bar pressing behavior ceased for several days as though there was an initial sensitization. After several weeks tolerance to this phenomenon was observed. A small dose (0.1 mg/kg) of LSD-25 given to DMT tolerant animals also showed some cross tolerance. Partial tolerance between LSD-25 and DMT was symmetric. Animals given LSD-25 daily became tolerant to this hallucinogen. When they were then given small doses of DMT disruption of bar pressing behavior was less than that observed in naive animals. Hence, it was concluded that tolerance occurs to the behavioral disruptive effects of DMT with partial cross tolerance to LSD (Kovacic and Domino, 1974). These findings are consistent with those of Rosenberg, Isbell, Miner, and Logan (1964) in which DMT was given to human subjects highly tolerant to LSD-25. Rosenberg et al. pointed out that except for the shorter onset and duration of action the peak effects of DMT in doses of 0.5 mg/kg in nontolerant humans were similar to those produced by 1.5 μg/kg of LSD-25. Subjects given daily LSD-25 to a maximum dose of 3.0 μg/kg twice a day showed marked tolerance. On the other hand, DMT, in a dose of 0.5 to 1.0 mg/kg given i.m., produced only a mild degree of cross tolerance to the mental response. There was no cross tolerance to the pupillary effects of DMT. The nature of their experimental design would have favored demonstration of a high degree of cross tolerance between LSD and DMT.

It is of interest that chronic schizophrenic patients usually do not respond to either LSD-25 or DMT as much as do normal subjects. This has often been interpreted as due to the fact that schizophrenic patients are apathetic and do not verbalize their feelings as readily. However, there are certain psychotropic drugs schizophrenic patients tolerate very poorly. For example, schizophrenic patients do not tolerate the psychotomimetic agent phencyclidine (Domino and Luby, 1973). Such patients are made considerably worse. It is well known that amphetamines may activate the schizophrenic process. Schizophrenic patients also do not tolerate MAO inhibitors (Ban, 1969), particularly tranylcypromine, especially if combined with TP and/or methionine (Alexander, Curtis, Sprince, & Crosley, 1963; Sprince, Parker, Jameson, & Alexander, 1963). Schizophrenic patients also are very sensitive to the administration of the dopamine β-hydroxylase and aldehyde dehydrogenase inhibitor, disulfiram (Antabuse) as reported by Heath, Vesselhof, Bishop, and Byers (1965). The fact that these latter drugs activate the schizophrenic process would suggest that the resistance of schizophrenics to LSD-25 and DMT could be interpreted as partial cross tolerance to endogenous psychotogenic substances. Tolerance to any psychoactive substance can be due to (a) behavioral mechanisms, (b) enhanced biotransformation or (c) an unknown cellular adaptive process. It would be of great interest to determine the pharmacokinetics of indole hallucinogens, especially their plasma half-life as well as their possible differential biotransformation in normals and schizophrenic patients. Szara (1961) and his associates (Szara and Axelrod, 1959; Szara, Hearst, & Putney, 1962) suggested that the biological effects of N-dialkyltryptamines may be due to biotransformation to corresponding 6-hydroxy metabolites. However, Holmstedt and Lindgren (1967) could find no evidence of such a pathway of metabolism. Furthermore, Rosenberg, Isbell, and Miner (1963) compared DMT and 6-hydroxy DMT given i.m. in doses of up to 1 mg/kg to former opiate addicts who had no major signs of physical illness or psychosis. It was quite apparent from their study that DMT provoked a marked subjective response with objective physiological changes while a similar dose of 6-hydroxy DMT produced only minimal actions. Thus, in man the 6-hydroxy metabolite hypothesis does not appear to be valid. It is generally known that hydroxyl substitutions produce compounds that penetrate the blood-brain barrier poorly as, for example, 5-HT and BFT in contrast to T and DMT. Nevertheless, additional research on the biotransformation of indole hallucinogens in animals and man and comparative metabolic studies in mentally normal and schizophrenic patients is essential.

Conclusion

Tolerance to DMT does occur, especially with low doses. However, this does not rule out a possible psychotogenic role of this agent in schizophrenia. Pharmacokinetic and metabolic studies of DMT and related hallucinogens seem essential.

WHAT ARE EFFECTIVE APPROACHES FOR PREVENTING THE SYNTHESIS, ENHANCING THE BREAKDOWN OR ANTAGONIZING THE EFFECTS OF HALLUCINOGENIC INDOLES?

Some of the possible therapeutic approaches to treating schizophrenic patients who produce hallucinogenic indoles is summarized below:

1. Low-TP diet + Nicotinamide.
2. Selective competitors with TP for AAAD.
3. Enhance TP oxgenase activity.
4. Selective inhibitors of NMT and INMT.
5. Compounds which trap methyl groups from SAM and MTHF, penetrate blood–brain barrier, and are pharmacologically inactive.

6. Enhance breakdown of indole hallucinogens.
   a. MAO
   b. Demethylation
   c. N-oxide formation
   d. Hydroxylation (?)
7. Antagonize effects of indole hallucinogens.

Inasmuch as TP is an essential amino acid that humans cannot biosynthesize in sufficient quantity, one approach could be to reduce TP intake in selected patients. One of the major problems is to reduce the TP content of the diet to a sufficient degree that would still be palatable and sufficiently nutritious to maintain life. Himwich (1971) has used this approach in the treatment of some schizophrenic patients. Unfortunately, his low-TP diet did not improve the patients. However, no blood TP measurements were taken in order to determine how effective the diet was. In addition, one would have to be certain to select schizophrenic patients on the basis that they produced hallucinogenic indoles, otherwise a low-TP diet would be of little value. The low-TP diet seems an especially rational approach in view of its analogy to the low-phenylalanine diet in the treatment of children with phenylketonuria. Of course, nicotinic acid supplements would have to be given to patients on a low-TP diet to prevent pellegra. It is of interest that many schizophrenic patients eat very poorly during an exacerbation of their illness. Whether this makes the mental disturbance worse or is beneficial to the patient is not known. It is a fact that acute schizophrenics have plasma-TP levels significantly below normal (Manowitz, Gilmour, Racevskis, 1973); Gilmour, Manowitz, Frosch, & Shopsin, 1973; Domino and Krause, 1974a). As the acute patients recover in the hospital, plasma-TP levels gradually return toward normal. Manowitz et al. and Gilmour et al. pointed out that acute schizophrenic patients had normal plasma phenylalanine and tyrosine levels at a time when plasma-TP was reduced. They concluded that the selective decrease in TP could not be due to starvation but they did not do a food-deprivation control to rule this out. It is well known that during periods of starvation and stress, such as in prisoners held in concentration camps or patients undergoing severe weight-reducing diets, there is an increased incidence of psychoses. Mentally normal people with severe dietary restrictions in the face of great stress would presumably have low plasma-TP levels but in addition probably would be deficient in many other important nutrients as well. Our present knowledge does not allow one to make a definitive conclusion as to whether selective low TP intake in schizophrenic patients would be beneficial therapy. Obviously, such a therapeutic approach would be relatively simple and could surely be justified on grounds of its potential merit. However, one must be certain that the patient is an indole hallucinogen producer and that blood-TP measurements be taken to insure a proper low TP intake.

As previously discussed, four major enzymes are involved in the biosynthesis of hallucinogenic indoles (see Figure 1). Aromatic amino acid decarboxylase (AAAD) is responsible for the decarboxylation of TP to T. Normally, very low levels of brain T are present but after a MAO inhibitor the levels of T increase 2- to 3-fold as mentioned earlier (Martin et al., 1972). AAAD is also involved in the decarboxylation of 5-HTP to 5-HT. There are much higher levels of 5-HT than T in the brain, so, apparently, 5-HTP is a much more effective substrate than TP for AAAD. The use of synthetic TP derivatives such as α-methyltryptophan would be an interesting approach to reduce brain levels of T and 5-HT. The rationale would be analogous to that of α-methyl DOPA (Aldomet) as a competing substrate for L-DOPA. α-Methyltryptophan has been shown to cause a prolonged decrease in brain 5-HT (Sourkes, 1971). A similar reduction in brain T would be expected. This agent also induces TP oxygenase activity in the liver which would further shunt TP away from the synthesis of hallucinogenic indoles. Selective 5-HTP supplements might be necessary to maintain brain 5-HT at reasonable levels. One of the disadvantages of α-methyltryptophan is that probably α-methyltryptamine would accumulate as a false neurotransmitter. Inasmuch as α-methyltryptamine is psychotoxic in both man (Murphree, Dippy, Jenney, & Pfeiffer, 1961) and animals (Vasko, Lutz, & Domino, 1974) the limiting factor to this therapeutic approach would be its accumulation in the brain.

Since T and 5-HT are substrates of both nonspecific NMT and the more specific INMT, another approach would be to find selective inhibitors of these enzymes. An important clue is the fact that SAH and DMT are product inhibitors of the NMT from rabbit lung (Krause and Domino, 1974). SAH is an especially potent inhibitor (K_i = 2.5 X 10^{-6} M) of the rabbit lung enzyme compared to DMT (K_i = 1.0 X 10^{-4} M) as was first shown by a medical student Ms. Bonnie Smith and a premedical student, Laurence Domino, working in my laboratory at the University of Michigan in the summer of 1972. Since then, Lin, Narasimhachari, and Himwich (1973) have described further the kinetics of SAH inhibition. It is known that SAH also inhibits a wide variety of methyltransferase reactions including phenethanolamine N-methyltransferase (PNMT) which catalyzes the methylation of norepinephrine to epinephrine (Deguchi & Barchas, 1972), catechol-O-methyltransferase (COMT) which methylates norepinephrine to normetanephrine (Deguchi & Barchas, 1972; Coward, D'Urso-Scott, & Sweet, 1972), acetylserotonin-O-methyltransferase (IOMT) which forms melatonin (Deguchi & Barchas, 1972), nicotinamide-N-methyltransferase which forms N-methyl-nicotinamide (Swiatek, Simon & Chao, 1973). SAM-homocysteine-S-methyltransferase which forms methionine (Shapiro, Almenas, & Thomson, 1965), glycine methyltransferase which forms sarcosine (Kerr, 1972), histamine-N-methyltransferase which forms N-methylhistamine (Zappia, Zydek-Cwick, & Schlenk, 1969), and tRNA methyltransferase which forms methylated polynucleotides (Kerr, 1972). Methylation of DNA which involves

SAM as the methyl donor has been suggested as playing a role in switching off genes when their role in development is over (Adams, 1973). SAH should be studied as an inhibitor of this methylation reaction.

In view of the lack of specificity of SAH as a methyltransferase inhibitor *in vitro*, it probably is fairly toxic *in vivo*. However I do not know of any data on the pharmacology and toxicology of SAH. There would be no great harm to the patient if PNMT and COMT were inhibited. Inhibition of IOMT would reduce melatonin levels which would not be too crucial for man and would be beneficial because 5-MODMT would not be formed. Since methionine is already in the diet, its formation *in vivo* from homocysteine does not seem too important nor does the *N*-methylation of histamine. The only systems which might be adversely affected by inhibition are the methylation of tRNA and DNA, but we need data to support this assumption. Even if SAH were active *in vivo* and not toxic one would still be concerned about its stability (for it is enzymatically broken down (Deguchi & Barchas, 1972) as well as its absorption and penetration through cell membranes and the blood–brain barrier. Furthermore, SAH is extremely expensive and it is unlikely that it can be tested in animals for toxicity and specificity of inhibition of methylation *in vivo* in the near future. However synthetic congeners of both SAH and DMT would be of considerable interest to test as potential selective NMT inhibitors, and if nontoxic, as therapeutic agents.

The use of nicotinamide in large doses as a treatment for schizophrenia was based upon its ability to trap excess methyl groups (Hoffer, Osmond, Callbeck, & Kahan, 1957). However, this substance does not lower SAM levels in the brain. This would be expected, for nicotinamide should not readily penetrate the blood–brain barrier because of its charged form at pH 7.4. The megavitamin approach using this substance has, unfortunately, not proved effective in the treatment of schizophrenia as reported by the recent American Psychiatric Association task force on megavitamins. Nevertheless, the approach of using alternate substrates for trapping methyl groups is still valid, as is finding more specific derivatives of SAH and related compounds that are effective NMT inhibitors.

Another approach for more effective treatment would be to enhance the breakdown of hallucinogenic indoles. Most of the hallucinogenic indoles are tertiary amines. Only T is a primary amine. The MAO enzymes are involved in their breakdown. Hence MAO inhibitors should enhance DMT effects. However, Sai-Halász (1963) showed that subjects given 100 mg of iproniazid a day for 4 days followed by a 2-day drug free period had remarkably few hallucinations and an attenuation of the DMT effect when given DMT on the 7th day. The volunteers did have an odd feeling of a changed personality following DMT which resembled the schizophrenic "wahnstimmung" which precedes the outbreak of a psychosis. In addition, Sai-Halász (1962a,b) showed that 1-methyl-D-lysergic acid butanolamide, a 5-HT antagonist, accentuated the experimental psychosis induced by DMT. This investigator felt that by elevating 5-HT brain levels one would protect against DMT hallucinations, while blocking 5-HT would enhance the effects of DMT. It is of interest that Wyatt, Vaughan, Galanter, Kaplan, and Green (1972) reported that the administration of 5-HTP to some schizophrenic patients had a beneficial effect. Unfortunately, this was not consistent or reproducible in all schizophrenic patients indicating a heterogeneity of disease types. There is an obvious discrepancy between the human studies of Sai-Halász and the findings from our own laboratory using animal models of DMT effects. For example, as illustrated in Figure 2 DMT-induced depression of rat shuttle-box acquisition is potentiated by iproniazid (Domino & Lutz, 1974). This is shown by a shift to the left in the dose-effect curve of DMT in suppressing one-way acquisition. Similarly, in rats trained for a FR_4 milk reward, single doses of iproniazid potentiate the suppression of DMT-induced barpressing (Kovacic and Domino, 1973). Pretreatment with iproniazid enhances rat brain

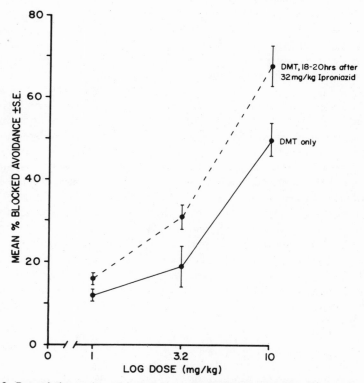

Figure 2. Potentiating action of iproniazid on DMT included suppression on rat one-way shuttle-box acquisition. Note shift of the dose-effect curve of DMT to the left after iproniazid pretreatment.

and liver levels of DMT (Lu, Wilson, Moore, & Domino, 1974). In addition, DMT-induced hyperthermia and pupillary dilatation in the rabbit is also prolonged by iproniazid (Moore, Demetriou, & Domino, 1974). From these animal findings it would appear that one pathway for the biodegradation of DMT is through MAO. *In vitro* DMT is a very poor substrate for MAO compared to T (Domino & Krause, 1974b). However, iproniazid *in vivo* appears to be far better in prolonging rat brain and liver levels of DMT than would be expected from *in vitro* studies (Lu et al., 1974).

These results, of course, do not rule out a prior demethylation of DMT to T before oxidation by MAO. This possibility is being pursued in our laboratory using demethylase inhibitors. If MAO is important for biotransforming DMT in man, it becomes most intriguing that platelet MAO activity is reduced in schizophrenic patients (Murphy & Wyatt, 1972; Wyatt, Murphy, Belmaker, Donnelly, Cohen, & Pollin, 1973; Domino & Sampath-Khanna, 1975). Although the mechanism by which platelet MAO activity is reduced in schizophrenic patients is unclear, it is sobering to note that brain MAO activity using tissue from deceased schizophrenic patients is not (Domino et al., 1973; Schwartz, Aikens, & Wyatt, 1974; Wise, Boden, & Stein, 1975). There are two basic families of brain MAO, the A and B types (Yang and Neff, 1974). In the studies quoted above, a nonspecific substrate T was used to determine MAO activity.

There are several other theoretical pathways by which DMT and related indole hallucinogens can be biotransformed. These include demethylation, N-oxide formation, and hydroxylation. In view of the essential role of the liver-microsomal enzymes in these metabolic reactions, it would be of considerable interest to study the effects of inhibitors of drug metabolism such as SKF 525A, or liver-microsomal enzyme inducers such as phenobarbital on the behavioral effects of DMT as well as on tissue levels of this substance. Perhaps by selectively enhancing the breakdown of DMT and related indoles one might be able to develop a novel treatment.

Another approach is to find more effective DMT antagonists. Although chlorpromazine antagonizes the hyperthermia, pupillary dilatation, and behavioral excitation of DMT in the rabbit, other neuroleptic agents with more selective anti-5-HT actions such as methiothepin are more effective (Moore et al., 1974). It is important to study other more selective antagonists of indole hallucinogens.

Conclusion

There are many novel new therapeutic approaches possible based upon the indole hallucinogen hypothesis. These seem worthy to pursue.

WHAT IS THE ROLE OF ANIMAL MODELS IN SEEKING NEW TREATMENTS OF SCHIZOPHRENIA BASED UPON THE INDOLE HALLUCINOGEN HYPOTHESIS?

If indole hallucinogens exist in some schizophrenic patients and are psychotogenic, then the use of animal models to find new treatments is of considerable value. Obviously, one of the major problems of finding an antagonist of a hallucinogen is to quantify hallucinations in animals. Yet this is impossible. No one can ever determine if an animal hallucinates! We can only study behavioral, chemical, and physiological end points. Let us consider some of the behavioral measures. They fall into two broad categories. (1) Those that are nonspecific, and (2) those that are specifically related to hallucinogens. Various nonspecific behavioral end points can be used to determine the disruptive effects of indole hallucinogens. One which we have used in our laboratory is one-way shuttle-box acquisition to an electric shock (Tenen, 1965; Caldwell, Oberleas, Clancy, & Praasad, 1970). DMT produces a dose-related suppression of avoidance acquisition in this task. The advantage of this behavior is that it is rapidly acquired; hence naive animals can be run the same day. Its disadvantage besides being nonspecific is it is rather insensitive to DMT. Another nonspecific task which lends itself to the study of the duration of action of hallucinogens is suppression of fixed-ratio responding for a food reward. We have used a FR_4 schedule for sweetened-milk reinforcement in rats food-deprived to maintain 70% of optimal weight. Such animals can easily be trained to bar press for food. The administration of DMT and related substances produces an abrupt cessation of operant responding and abrupt recovery. Hence, the duration of action of the hallucinogen can be measured as the period of nonresponding. A large variety of other operant behaviors could also be used. However, drugs which are potential antagonists must in themselves not affect such behaviors otherwise they cannot be tested as suitable antagonists. One of the problems highlighted by the administration of a potential antagonist is that in the two behaviors used in our laboratory, chlorpromazine has only been able to potentiate the actions of d-amphetamine and not antagonize them as is generally known for other behaviors. However, this may simply be a dose-related phenomenon for small doses of chlorpromazine prolong amphetamine-induced stereotypes as well as other behavioral effects (Borella, Herr, & Wojdan, 1969; Lal & Sourkes, 1972). Inasmuch as larger doses of chlorpromazine reduce behavioral responding, nonspecific behaviors are very limited in testing for antagonists of hallucinogenic activity. Appel (1968) has reviewed extensively the effects of psychotomimetic drugs, especially LSD-25, on animal behavior. Other behavioral models using animals to which he has referred can probably be very useful to screen potential antagonists.

A biochemical measure which can be used to test for indole hallucinogens of the LSD-25 type is an apparent reduction in brain 5-HT turnover. Brain 5-HT increases and 5-HIAA decreases following LSD and related compounds (Aghajanian & Freedman, 1968). Inasmuch as MAO inhibitors do the same, one must be certain MAO inhibition is not involved. Hence the end point is indirect but useful.

Other physiologic end points such as temperature elevation and pupillary dilatation can also be used for assaying hallucinogenic activity. It is well known that hallucinogens of the LSD-25 type elevate the rectal temperature of rabbits (Rothlin, 1957). This is illustrated in Figure 3 which compares the effects of 10 µg/kg of LSD-25 with 3.2 mg/kg of DMT given i.v. Obviously, the data of this figure are of limited value inasmuch as dose-effect comparisons of both agents are needed. This is now being done in our laboratory. Although hyperthermia is a nonspecific end point, it can be used for determining potential antagonistic effects. In this test, both chlorpromazine and methiothepin are effective antagonists of DMT-induced hypothermia and pupillary dilatation (Moore et al., 1974). Again, this test suffers from a lack of specificity inasmuch as psychomotor stimulants also elevate rectal temperature. Furthermore, with α-methyl-p-tyrosine one can dissociate LSD-induced increase in rectal

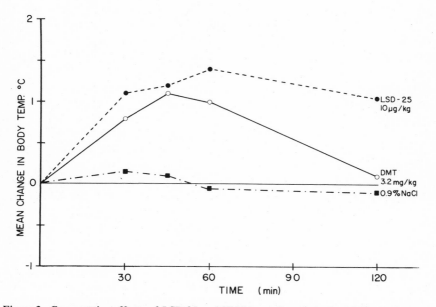

Figure 3. Comparative effects of LSD-25 and DMT hyperthermia in the rabbit. Both agents were given iv. Each point represents the mean of 3–6 rabbits.

temperature from its other behavioral effects which are reduced (Horita & Hamilton, 1969).

More specific tests of hallucinogenic activity include head bobbing in mice and splayed legs in rodents or splayed wings in chicks. To my knowledge these end points have not been used for studying hallucinogen antagonists. However, T convulsions have been used as an end point in rats for studying potential 5-HT antagonists (Tedeschi, Tedeschi, & Fellows, 1959). Bovet–Gatti profiles based upon the distribution of premature, optimal, or late responses on rat bar-pressing behavior to avoid electroshock have also been suggested as an excellent way of screening for hallucinogens (Bovet & Gatti, 1965; Smythies, 1970) although we have been disappointed with this approach. One can also use various neurophysiologic end points for screening potential antagonists of hallucinogens such as LSD-25 (Koella, 1968). Inasmuch as LSD-25 and indole hallucinogens block the lateral geniculate-evoked response (Bishop, Field, Hennessy, & Smith, 1958, 1959; Evarts, Landau, Freygang, & Marshall, 1955; Evarts, 1958) one could use this response as a test of potential antagonists, using the cat or monkey with acute or chronic indwelling brain electrodes. This approach suffers in that it is not a suitable rapid-screening procedure and either acute animals must be used under general anesthesia or chronic animals who move about and hence are likely to show more variability. Another approach with similar limitations is to measure the rate of firing of the serotonergic neurons in the raphé nucleus. After LSD-25 and mescaline the firing rates are reduced (Aghajanian, Sheard, & Foote, 1970). These investigators have also shown that DMT reduces raphé-neuron firing.

CONCLUSION

As yet there are no satisfactory animal models for studying indole hallucinogen antagonists. Most are either nonspecific or very time consuming. Should DMT and related agents be shown to be psychotogenic even in a minority of schizophrenics, it will behoove us all to use such animal models to a maximum. At present our animal models are all that we have to find new treatments for our mentally ill. After all, much of the progress of modern medicine came about in our lifetime because researchers were able to reproduce human disease in animal models and hence seek new ways of treatment.

ACKNOWLEDGMENTS

The author would like to acknowledge a most remarkable psychiatrist and friend, Dr. Jacques Gottlieb, who made the research reported possible at the

Lafayette Clinic. We hope the ideas expressed will provide some better treatments. In addition, the personnel of the pharmacology laboratory whose research is described have provided the data upon which this manuscript is based. This study was supported by an appropriation from the State of Michigan to the Lafayette Clinic for the study of schizophrenia under the direction of Dr. J. S. Gottlieb.

REFERENCES

Adams, R. L. P. Delayed methylation of DNA in developing sea urchin embryos. *Nature (New Biol.)*, 1973, *244:131*, 27–29.

Aghajanian, G. K., & Freedman, D. X. Biochemical and morphological aspects of LSD pharmacology. In D. H. Efron (Ed.), *Psychopharmacology—A Review of Progress 1957–1967*, PHSP #1836. Superintendent of Documents, Washington, D.C., 1968, pp. 1185–1193.

Aghajanian, G. K., Sheard, M. H., & Foote, W. E. LSD and mescaline: comparison of effects on single units in the midbrain raphé. In D. H. Efron (Ed.), Psychotomimetic Drugs, New York: Raven Press, 1970, pp. 165–170.

Ahn, H. A., Walker, R. W., VandenHeuvel, W. J. A., Rosegay, A., & Mandel, L. R. Studies on the *in vivo* biosynthesis of *N,N*-dimethyltryptamine (DMT) in the rabbit and rat. *Fed. Proc.* 1973, *32*, 511.

Alexander, F., Curtis, G. C., III, Sprince, H. & Crosley, A. P., Jr. L-Methionine and L-tryptophan feedings in non-psychotic and schizophrenic patients with and without tranylcypromine. *J. Nerv. Ment. Dis.*, 1963, *137*, 135–142.

Appel, J. B. The effects of "psychotomimetic" drugs on animal behavior. In D. H. Efron (Ed.), Pharmacology—A Review of Progress 1957–1967, PHSP #1836, Superintendent of Documents, Washington, D.C., 1968, pp. 1211–1222.

Axelrod, J. The enzymatic *N*-methylation of serotonin and other amines. *J. Pharmacol. Exp. Ther.* 1962, *138*, 28–33.

Ban, T. Monoamine oxidase inhibitor antidepressants, In *Psychopharmacology*. Baltimore: Williams and Wilkins, 1969.

Banarjee, S. P. & Snyder, S. H. Methyltetrahydrofolic acid mediates *N*- and *O*-methylation of biogenic amines. *Science* 1973, *182*, 74–75.

Barchas, J. D., Elliott, G. R., DoAmaral, J., Erdelyi, E., O'Connor, S., Bowden, M., Brodie, H. K. H., Berger, P. A., Renson, J., & Wyatt, R. J. Triptolines: Formation from tryptamines and 5-MTHF by human platelets. *Arch. Gen. Psychiat.* 1974, *31*, 862–867.

Bidder, T. G., Mandel, L. R., Ahn, H. S., Walker, R. W., & VandenHeuvel, W. J. A. Blood and urinary DMT concentrations in acute psychotic disorders. *Lancet*, 1974, *1*, 165.

Bishop, P. O., Burke, W., & Hayhow, W. R. Lysergic acid diethylamide block of lateral geniculate synapses and relief by repetitive stimulation. *Exp. Neurol.* 1959, *1*, 556–568.

Bishop, P. O., Field, G., Hennessy, B. L., & Smith, J. R. Actions of D-lysergic acid diethylamide on lateral geniculate synapses. *J. Neurophysiol.*, 1958, *21*, 529–549.

Borella, L., Herr, F., & Wojdan, A. Prolongation of certain effects of amphetamine by chlorpromazine. *Canad. J. Physiol. and Pharmacol.*, 1969, *47*, 7–13.

Bovet, D., & Gatti, G. L. Pharmacology of instrumental avoidance conditioning. In Proceedings of 2nd International Pharmacological Meeting, Prague, 1965, pp. 75–89.

Caldwell, D. F., Oberleas, D., Clancy, J. J., & Praasad, A. A. Behavioral impairment in adult rats following acute zinc deficiency. *Proc. Soc. Exp. Biol. Med.*, 1970, *133*, 1417–1421.

Cohen, I. & Vogel, W. H. Determination and physiological disposition of dimethyltryptamine in rat brain, liver and plasma. *Biochem. Pharmacol.,* 1972, *21*, 1214–1218.

Coward, J. K., D'Urso-Scott, M., & Sweet, W. D. Inhibition of catechol-*O*-methyltransferase by *S*-adenosylhomocysteine and *S*-adenosylhomocysteine sulfoxide, a potential transition-state analog. *Biochem. Pharmacol.,* 1972, *21*, 1200–1203.

Deguchi, T., & Barchas, J. Inhibitions of transmethylations of biogenic amines by *S*-adenosylhomocysteine. *J. Biol. Chem.,* 1972. *246*, 3175–3181.

Domino, E. F., Krause, R. R., & Bowers, J. Various enzymes involved with putative neurotransmitters—regional distribution in the brain of deceased mentally normal, chronic schizophrenics or organic brain syndrome patients. *Arch. Gen. Psychiat.,* 1973, *29*, 195–201.

Domino, E. F., & Krause, R. R. Plasma tryptophan tolerance curves in drug free normal controls, schizophrenic patients and prisoner volunteers. *J. Psychiat. Res.,* 1974a, *10*, 247–261.

Domino, E. F., & Krause, R. R. Unpublished observations, 1974b.

Domino, E. F., & Luby, E. D. Abnormal mental states induced by phencyclidine as a model of schizophrenia. In J. O. Cole, A. Friedhoff, and A. M. Friedman (Eds.), *Psychopathology and Psychopharmacology,* Proc. 62 Meeting of Amer. Psychopath. Assoc., Baltimore: Johns Hopkins Press, 1973, pp. 35–50.

Domino, E. F., & Lutz, M. F. Unpublished observations, 1974.

Domino, E. F., & Sampath-Khanna, S. Decreased platelet MAO in chronic schizophrenics. *Amer. J. Psychiat,* 1975, in press.

Evarts, E. V. Effects of a series of indoles on synaptic transmission in the lateral geniculate nucleus of the cat. In H. H. Pennes (Ed.), Progress in Neurobiology, III—Psychopharmacology, New York, Hoeber-Harper, 1958, pp. 173–194.

Evarts, E. V., Landau, W., Freygang, W., Jr., & Marshall, W. H. Some effects of lysergic acid diethylamide and bufotenine on electrical activity in the cat's visual system. *Am. J. Physiol.,* 1955. *182*, 594–598.

Fisher, R. Chemistry of the brain. *Nature* 1968, *220*, 411.

Franzen, Fr., & Gross, H. Tryptamine, *N,N*-dimethyltryptamine, *N,N*-dimethyl-5-hydroxytryptamine and 5-methoxytryptamine in human blood and urine. *Nature,* 1965, *206*, 1052.

Gillin, J. C., Cannon, E., Magyar, R., Schwartz, M., & Wyatt, R. J. Failure of DMT to evoke tolerance in cats. *Biol. Psychiat.,* 1973, *7*, 213–220.

Gilmour, D. C., Manowitz, P., Frosch, W. A., & Shopsin, B. Association of plasma tryptophan levels with clinical change in female schizophrenic patients. *Biol. Psychiat.,* 1973, *6*, 119.

Greenberg, R. *N,N*-Dimethylated and *N,N*-diethylated indoleamines in schizophrenia. In H. C. Sabelli (Ed.), *Chemical Modulation of Brain Function—A tribute to J. E. P. Toman,* New York, Raven Press, 1973, pp. 277–296.

Gross, V. H., & Franzen, Fr. Zur bestimmung korpereigener amine in biologischen substraten. *Zeitschrift für Klinische Chemie,* 1965, *3*, 99–102.

Heath, R. G., Vesselhof, W., Bishop, N. P., & Byers, L. W. Behavioral and metabolic changes associated with administration of tetraethylthiuram disulfide. *Dis. Nerv. System* 1965, *26*, 99–106.

Himwich, H. E. (Ed.) *Biochemistry, Schizophrenias and Affective Illnesses,* Baltimore: Williams and Wilkins, 1971.

Himwich, H. E., Narashimhachari, N., Heller, B., Spaide, J., Haskovec, L., Fujimori, M., & Tabushi, K. Biochemical approaches to the study of schizophrenia. In H. C. Sabelli (Ed.), *Chemical Modulations of Brain Function—A Tribute to J. E. P. Toman,* New York: Raven Press, 1973.

Hoffer, A., Osmond, H., Callbeck, M. J., & Kahan, I. Treatment of schizophrenia with nicotinic acid and nicotinamide. *J. Clin. Exp. Psychopath.*, 1957, *18*, 131–158.

Holmstedt, B., & Lindgren, J. B. Chemical constituents and pharmacology of South American snuffs. In D. H. Efron (Ed.), *Ethnopharmacologic Search for Psychoactive Drugs*, Publication No. 1645, Public Health Service, Washington, D.C., 1967, pp. 339–382.

Horita, A., & Hamilton, A. E. Lysergic acid diethylamide: Dissociation of its behavioral and hyperthermic actions by DL-alpha-methyl-*p*-tyrosine. *Science*, 1969, *164*, 78.

Hsu, L. L., & Mandell, A. J. Multiple *N*-methyltransferases for aromatic alkylamines in brain. *Life Sci.*, 1973, *13*, 847–868.

Isbell, H. Quoted in D. H. Efron (Ed.), *Ethnopharmacologic Search for Psychoactive Drugs*, Publication No. 1645, Public Health Service, Washington, D. C., 1967, p. 369.

Juntunen, J. M., Struck, W. G., Warner, K. A., Frohman, C. E., & Gottlieb, J. S. Effects of tryptophan loading on indole metabolite levels in blood and urine of schizophrenic patients. *Biol. Psychiat.*, 1974, in press.

Kaplan, J., Mandel, L. R., Stillman, R., Walker, R. W., VandenHeuvel, W. J. A., Gillin, J. C. & Wyatt, R. J. Blood and urine levels of *N*,*N*-dimethyltryptamine following administration of psychoactive dosages to human subjects. *Psychopharmacologia*, 1974, *38*, 239–245.

Kerr, S. J. Competing methyltransferase systems. *J. Biol. Chem.* 1972, *247*, 4248–4252.

Koella, W. P. Neurophysiological effects of psychotomimetic substances (A supplemental review). In D. H. Efron (Ed.), *Psychopharmacology—A Review of Progress 1957–1967*, PHSP #1836, Superintendent of Documents, Washington, D.C., 1968, pp. 1223–1230.

Koella, W. P., Beaulieu, R. F., & Bergen, J. R. Stereotyped behavior and cyclic changes in response produced by LSD in goats. *Int. J. Neuropharmacol.*, 1964, *3*, 397–403.

Kovacic, B., & Domino, D. F. Effects of various drugs on DMT-induced disruption of rat bar pressing behavior. *The Pharmacologist*, 1973, *15*, 218.

Kovacic, B. & Domino, E. F. Tolerance to behavioral effects of dimethyltryptamine (DMT) in the rat. *Fed. Prod.*, 1974, *33*, 549.

Krause, R. & Domino, E. F. Kinetics of product inhibition of rabbit lung *N*-methyltransferase. *Res. Comm. Chem. Path. and Pharmacol.* 1974, *9:2*, 399–373.

Lal, S., & Sourkes, T. L. Effect of various chlorpromazine metabolites in amphetamine-induced stereotyped behavior in the rat. *Europ. J. Pharmacol.*, 1972. *17*, 283–286.

Lin, R. L., Narasimhachari, N., & Himwich, H. E. Inhibition of indolethylamine-*N*-methyltransferase by *S*-adenosylhomocysteine. *Biochem. Biophys. Res. Comm.* 1973, *54*, 751–759.

Lipinski, J., Mandel, L. R., Ahn, H. S., VandenHeuvel, W. J. A., & Walker, R. W. Blood dimethyltryptamine concentrations in psychotic disorders. *Biol. Psychiat.*, 1974, *9:1*, 89–91.

Lu, L. W., Wilson, A., Moore, R. H., & Domino, E. F. Correlation between brain *N*,*N*-dimethyltryptamine (DMT) levels and bar pressing behavior in rats: Effect of MAO inhibition. *The Pharmacologist*, 1974, *16*, 237.

Mandel, L. R. Dimethyltryptamine: Its biosynthesis and possible role in mental disease. In E. F. Domino and J. M. Dau (Eds.), *Neurotransmitters Balances Regulating Behavior*. Ann Arbor, Michigan, Edwards Bros., Inc., 1975, in press.

Mandel, L. R., Rosegay, A., Walker, R. W., VandenHeuvel, W. J. A., & Rokach, J. 5-Methyltetrahydrofolate acid as a mediator in the formation of pyridoindoles. *Science*, 1974, *186*, 741–743.

Mandel, L. R., Ahn, H. S., VandenHeuvel, W. J. A., & Walker, R. W. Indoleamine-*N*-methyltransferase in human lung. *Biochem. Pharmacol.*, 1972, *21*, 1197–1200.

Mandel, L. R., Rosenzweig, S., & Kuehl, F. A., Jr. Purification and properties of indoleamine-*N*-methyltransferase. *Biochem. Pharmacol.*, 1971, *20*, 712–716.

Manowitz, P., Gilmour, D. G., & Racevskis, J. Low plasma tryptophan levels in recently hospitalized schizophrenics. *Biol. Psychiat.*, 1973, *6*, 109–118.

Martin, W. R., Sloan, J. W., Christian, S. T., & Clements, T. H. Brain levels of tryptamine. *Psychopharmacologia (Berl.)*, 1972, *24*, 331–346.

Moore, R. H., Demetriou, S., & Domino, E. F. Effects of iproniazid, chlorpromazine and methiothepin on DMT-induced changes in body temperature, pupil dilatation, blood pressure and EEG in the rabbit. *Arch. Int. Pharmacol. et Dynamie., et de Ther.*, 1975, *213*, 64–72.

Murphree, H. B., Dippy, R. H., Jenney, E. H., & Pfeiffer, C. C. Effects in normal man of α-methyltryptamine and α-ethyltryptamine. *Clin Pharmacol. Ther.*, 1961, *2*, 722–726.

Murphy, D. L., & Wyatt, R. J. Reduced monoamine oxidase activity in blood platelets from schizophrenic patients. *Nature* 1972, *238*, 225–226.

Narasimhachari, N., Heller, B., Spaide, J., Haskovec, L., Fujimori, M., Tabushi, K., & Himwich, H. E. Urinary studies of schizophrenics and controls. *Biol. Psychiat.*, 1971a, *3*, 9–20.

Narasimhachari, N., Heller, B., Spaide, J., Haskovec, L., Meltzer, H., Strahilevitz, M., & Himwich, H. E. *N-N*-Dimethylated indoleamines in blood. *Biol. Psychiat.*, 1971b. *3*, 21–23.

Narasimhachari, N., Plaut, J. M., & Himwich, H. E. Indole ethylamine-*N*-methyltransferase in serum samples of schizophrenics and normal controls. *Life Sci.*, 1972, *11*, 221–227.

Rosenberg, D. E. Isbell, H., & Miner, E. J. Comparison of a placebo, *N*-dimethyltryptamine, and 6-hydroxy-*N*-dimethyltryptamine in man. *Psychopharmacologia (Berl.)*, 1963, *4*, 39–42.

Rosenberg, D. E., Isbell, H., Miner, E. J., & Logan, C. R. The effect of *N,N*-dimethyltryptamine in human subjects tolerant to lysergic acid diethylamide. *Psychopharmacologia (Berl.)*, 1964, *5*, 217–227.

Rosengarten, H., Szemis, A., Piotrowski, A., Romaszewska, K., Matsumato, H., Stencka, K., & Jus, A. *N, N*-Dimethyltryptamine and bufotenin in the urine of patients with chronic and acute schizophrenic psychoses. *Psychiatria Polska*, 1970, *4*, 519–521.

Rothlin, E. Pharmacology of lysergic acid diethylamide and some of its related compounds. In S. Garattini and V. Ghetti (Eds.), *Psychotropic Drugs*, Amsterdam: Elsevier, 1957, pp. 36–47.

Saavedra, J. M., & Axelrod, J. Psychotomimetic *N*-methylated tryptamines: formation in brain *in vivo* and *in vitro. Science*, 1972, *172*, 1365–1366.

Saavedra, J., Coyle, J. R., & Axelrod, J. The distribution and properties of the non-specific *N*-methyltransferase in brain. *J. Neurochem.*, 1973, *20*, 743–752.

Sai-Halász, A. The effect of antiserotonin on the experimental psychosis induced by dimethyltryptamine. *Experentia (Basel)*, 1962a, *18*, 137–138.

Sai-Halász, A. The role of serotonin antagonism in the model-psychoses. *(Hung.) Iddeggyog-Szle*, 1962b, *15*, 301.

Sai-Halász, A. The effect of MAO inhibition on the experimental psychosis induced by dimethyltryptamine. *Psychopharmacologia (Berl.)*, 1963, *4*, 385–388.

Schwartz, M. A., Aikens, A. M., & Wyatt, R. J. Monoamine oxidase activity in brains from schizophrenic and mentally normal individuals. *Psychopharmacologia*, 1974, *38*, 319–328.

Seevers, M. H., & Deneau, G. A. Physiological aspects of tolerance and physical dependence. In W. S. Root and F. G. Hofmann (Eds.), *Physiological Pharmacology*, Vol. 1, Part A, New York: Academic Press, 1963, pp. 565–640.

Shapiro, S. K., Almenas, A., & Thomson, J. F. Kinetics and mechanism of reaction of *S*-adenosylmethionine: homocysteine methyltransferase. *J. Biol. Chem.*, 1965, *240*, 2512–2518.

Smythies, J. R. (Ed.). The Mode of Action of Psychotomimetic Drugs. A Report Based On An NRP Work Session. Neurosciences Research Program Bulletin, 1970, *8, 1*, 152.

Sourkes, T. L. Alpha-methyltryptophan and its actions on tryptophan metabolism. *Fed. Proc.*, 1971, *30*, 897–903.

Sprince, H., Parker, C. M., Jameson, D., & Alexander, F. Urinary indoles in schizophrenic and psychoneurotic patients after administration of tranylcypromine (Parnate) and methionine or tryptophan. *J. Nerv. Ment. Dis.*, 1963, *137*, 246–251.

Swiatek, K. R., Simon, L. N., & Chao, K.-L. Nicotinamide methyltransferase and S-adenosylmethionine: 5[1]-methylthioadenosine hydrolase. Control of transfer ribonucleic acid methylation. *Biochem.*, 1973, *12*, 4670–4674.

Szara, S. Hallucinogenic effects and metabolism of tryptamine derivatives in man. *Fed. Proc.*, 1961, *20*, 285–288.

Szara, S., & Axelrod, J. Hydroxylation and N-demethylation of N,N-dimethyltryptamine. *Experientia*, 1959, *15*, 216–217.

Szara, S., Hearst, E., & Putney, F. Metabolism and behavioral action of psychotropic tryptamine homologues. *Int. J. Neuropharmacol.*, 1962, *1*, 111–117.

Tedeschi, D. H., Tedeschi, R. E., & Fellows, E. J. The effects of tryptamine on the central nervous system, including a pharmacological procedure for the evaluation of iproniazid-like drugs. *J. Pharmacol. Exp. Ther.*, 1959, *126*, 223–232.

Tenen, S. S. An automated one-way avoidance box for the rat. *Psychonom. Sci.*, 1965, *6*, 407–408.

Vasko, M., Lutz, M. P., & Domino, E. F. Structure activity relations of some idolealkyamines in comparison to phenethylamines on motor activity and acquisition of avoidance behavior. *Psychopharmacologia*, 1974, *36*, 49–58.

Walker, R. W., Ahn, H. S., Albers-Schonberg, G., Mandel, L. R., & VandenHeuvel, W. J. A. Gas chromatographic-mass spectrometric isotope dilution assay for N,N-dimethyltryptamine in human plasma. *Biochem. Med.*, 1973, *8*, 105–113.

Wise, C. D., Baden, M. M., & Stein, L. Postmortem measurement of enzymes in human brain: Evidence of a central noradrenergic deficit in schizophrenia. In E. F. Domino and J. Davis (Eds.), *Neurotransmitter Balances Regulating Behavior*, Ann Arbor, Michigan, Edwards Bros, Inc., 1975, in press.

Wyatt, R. J., Vaughan, T., Galanter, M., Kaplan, J., & Green, R. Behavioral changes of chronic schizophrenic patients given L-5 hydroxytryptophan. *Science*, 1972, *177*, 1124–1126.

Wyatt, R. J., Mandel, L. R., Ahn, H. S., Walker, R. W. & VandenHeuvel, W. J. A. Gas chromatographic-mass spectrometric isotope dilution determination of N,N-dimethyltryptamine concentrations in normals and psychiatric patients. *Psychopharmacologia (Berl.)*, 1973a, *31*, 265–270.

Wyatt, R. J., Mandel, L. R., Ahn, H. S., Walker, R. W. & VandenHeuvel, W. J. A. DMT—a possible relationship to schizophrenia. Proc. 1st Intl. Serotonin Conference. 1973b, in press.

Wyatt, R. J., Murphy, D. L., Belmaker, R., Donnelly, D., Cohen, S., & Pollin, W. Reduced monoamine oxidase activity in platelets: a possible genetic marker for vulnerability to schizophrenia. *Science*, 1973c, *179*, 916–918.

Wyatt, R. J., Saavedra, J. M., & Axelrod, J. A dimethyltrptamine (DMT) forming enzyme in human blood. *Amer. J. Psychiat.*, 1973d, *130*, 754–760.

Yang, H. Y. T., & Neff, N. H. The monoamine oxidases of brain: selective inhibition with drugs and the consequences of the metabolism of the biogenic amines. *J. Pharmacol. Esp. Ther.*, 1974, *189*, *3*, 733–740.

Zappia, V., Zydek-Cwick, C. R., & Schlenk, F. The specificity of S-adenosylmethionine derivatives in methyl transfer reactions. *J. Biol. Chem.*, 1969, *244*, 4499–4509.

Discussion

ARTHUR KLING

Since there has already been considerable discussion of the formal presentations, I will forego commenting on each paper and raise some general issues not yet dealt with in any depth.

With respect to the use of animal models relating neural function to behavior, much of what we know about the nervous system is based on studies conducted in animals. Work on single nerves or neurons to complex interactions in sensory and motor systems are all essentially based on animal models which have been invaluable in our understanding of the human nervous system. When we go beyond the physiological into the behavioral area, the waters get a bit muddy.

In some of the presentations, terminology has been used to describe behavioral syndromes in animals which are used in clinical psychiatry to describe human disorders. The problem of nosology in clinical psychiatry has hardly been solved and is still beset with major problems of terminology surrounding human disorders. We should be very cautious in the application of clinical terminology to infrahuman subjects. Vague, misapplied or confusing names for human syndromes have been responsible for considerable difficulty in interpreting results of research and treatment of human disorders and, in many cases, have led to the misapplication or research findings and perpetration of inaccurate treatments.

Good research should not be clouded by the use of terms which could be confusing and ultimately interfere with the understanding of the behavioral determinants. Words such as *depression, despair, phobias*, and *hallucinations* have been used to describe various behavioral states in animals. Psychiatry is still having trouble with the accurate use of these terms with respect to clinical prob-

ARTHUR KLING ● College of Medicine and Dentistry of New Jersey, Rutgers Medical School, Piscataway, New Jersey.

lems, let's not compound the problems by applying these words to animal species. Dr. Harlow's group and others working in the area of separation and affiliation have made highly significant contributions to fundamental notions about attachment processes, mother–infant and peer interactions. Why do we need to talk about these behavioral abnormalities as though they were human clinical syndromes? It may turn out, for example, that the abnormal behaviors resulting from social isolation have little to do with most human depressive syndromes and, therefore, we may overlook the significant aspects because the "label" has directed out attention elsewhere. Even in cross-cultural studies there is considerable confusion surrounding the use of western diagnostic categories as applied to behavioral syndromes seen in other societies.

I want to now talk about an area which has been touched upon at this conference but which needs some additional comment; that is, how we observe animals and how environment interacts with experimental variables to alter the consequences of a manipulation. Neurophysiological studies as applied to behavior, traditionally have been done in the laboratory under very controlled conditions. And all of us who have trained in that area respect the use of certain boundaries to reduce the variability of our experiments. However, when it comes to behavior, I think we have to take a look at what the ethologists and anthropologists have been doing for many years, to learn from them something about how to observe animals and the importance of the setting in which these behaviors occur. This is where we have been directing our attention for the past few years. And I would like to show you one example of how this applies to brain-behavior research. Some four decades ago, Drs. Kluever and Bucy published their now classical report on the behavior of monkeys subjected to bitemporal lobectomy. They described a syndrome consisting of hyperorality, hypersexuality; that is, the monkey mounts inappropriate objects and shows excessive autoerotic activity, it becomes rather tame toward man, shows hypermetamorphosis and a visual agnosia. This syndrome now has been seen in many laboratories, and by many different investigators. With few exceptions, all of these observations, however, have been made on animals observed in a laboratory cage. In some cases, these subjects were transferred to a test apparatus in which they were presented with cognitive problems. Many years ago, Klüever remarked that it is a paradox to him as to how a monkey who behaves so bizarrely could, under test situations, learn very complex discriminations. We still do not understand it, but we do have interesting findings which bear directly upon this problem. What we have been doing is trying to study the effects of amygdaloid (and other) ablations in primates in a social context; that is, to what extent does this ablation affect their social behavior. Our usual laboratory group is composed of ferral-born stumptail macaques, numbering between nine and eleven or so, which are placed in an enclosure and their interactions are observed over a period of time, and quantitatively recorded. Following this, animals are removed and operated and then returned to the enclosure. If we, for example, look at just one of the behaviors, joining, for example, you will note that there is an increase in joining among all the subjects in this enclosure over a period of three months, representing approximately 200 hours of quantitative observation preoperatively and the same postoperatively. The mean frequencies between the operates and the controls do not look very different. We then removed this group and placed them in a half-acre enclosure at the Caribbean Primate Research Center in Puerto Rico. Now you can see there is considerable difference between the rates of joining between the operates and the normals. That is, the operates now show few instances of joining and this persists over time. If we examine grooming, the operated subjects show a decrease in the laboratory while the controls show an increase over time. In the corral, there is a further decrement, the operates are hardly observed to groom at all, particularly with normals. If we look at their sexual behavior, in the laboratory enclosure, we see a remarkable range of sexual behaviors; all the way from masturbation, autofellatio, heteroxsexual masturbation, heterosexual genital licking, heterosexual fellatio, homosexual masturbation—female and male—and, finally, copulation. As you can see, again, there is marked qualitative and quantitative difference between preoperative and postoperative behaviors. The important thing to note is that it is not the operates alone who have changed. That is, the normals in the group are also engaged in these behaviors with them. In the corral setting these behaviors disappear.

When we study monkeys out of the corral, in a totally natural setting, the lesioned subjects become social isolates. That is, they do not interact with their conspecifics. We can now, from a number of experiments, show that this decrement in social behavior is directly related to the setting and to the social complexity in which the animals are living.

In 1968, Jane Lancaster and I studied the effects of amygdala lesions in totally free-ranging African green monkeys. Very briefly, unoperated animals who were trapped, kept in captivity and released, all wound up either in their own band or within another group within seven to ten days after release. In contrast, none of the operated subjects, whether it was the dominant male of the group or an adult female, or a juvenile, returned to a group. All remained social isolates and none ever showed any of the symptoms commonly seen after this lesion in laboratory housed subjects. We have, then, a well-known syndrome, easily reproducible in the laboratory which shows not only dramatic alteration depending on the setting and group composition, but also a new aspect; namely, social indifference and isolation when the monkeys are in their natural habitat. Time does not allow a discussion of how environment affects the behavior of nonhuman primates subjected to other brain lesions, suffice it to say that utilizing more natural settings for behavioral observation is providing new insights into brain-behavior mechanisms particularly with respect to affiliative and aggressive behaviors and that we should be cautious in overgeneralizing the results of laboratory studies.

Animal Models for Human Psychopathology
Observations from the Vantage Point of Clinical Psychopharmacology

DENNIS L. MURPHY

In our clinical studies of depressed and manic individuals, we regularly make use of data from animal studies in planning our research and in evaluating our results. The data we use most come from studies of the neurochemical effects of drugs, and, to a lesser extent, data on the effects of drugs on the behavior of normal animals. In contrast, we rarely use biochemical or behavioral data derived from studies of abnormal animals—either strains exhibiting altered behavior or neuro-chemistry, or animals with induced "psychopathology," because very little such information is available. One rapidly growing area of investigation, however, which was discussed in papers by Drs. Domino and Irwin, is the utilization of drugs which produce states of abnormal behavior in animals and man. Some of the more prominent examples of these pharmacologic agents are listed in Table I.

The limited availability of animal models for psychopathologic states is of concern because recent and still rather limited information from studies of the major psychoactive drugs that we use clinically, including the tricyclic anti-depressants and lithium, indicates that these drugs have very minimal behavioral

DENNIS L. MURPHY ● Laboratory of Clinical Science, National Institute of Mental Health, Bethesda, Maryland.

Table I. Some Pharmacologic Agents Producing Model Psychopathologic Syndromes

Schizophrenia:	Amphetamine, 6-hydroxydopamine, LSD, DMT, phencyclidine
Depression:	Reserpine, α-methyl tyrosine, physostigmine
Mania:	Amphetamine, morphine, desipramine-tetrabenazine, MAO inhibitors
Anxiety:	Lactic acid

effects in normal human individuals, despite their marked effects in patients with affective disorders. For instance, the tricyclic drugs are not euphoriants in normals. Also, drugs such as amphetamine, which regularly yield marked activation and euphoria in normal individuals, do not lead to sustained antidepressant effects in patients. Normal people or normal animals appear not to be the appropriate testing ground for studies of the most effective groups of psychoactive drugs, and there is a great need for valid animal models for the affective disorders and other psychopathologic processes.

In fact, our needs are somewhat different from those described for behavioral science in general by Dr. Hinde. We would very much like to have multiple, overlapping models—each of which would teach us something different in its similarities to and differences from the human state. Taking into due consideration the cautions mentioned by Dr. Hinde, there would be dividends from having a valid animal model with the maximum similarity possible to the human disorders—essentially a simplified replica of the disorders. Ideally, this model would have similar phenomenology, perhaps a similar etiology, and would respond to similar treatments as are effective in the human condition. Such a model would be especially useful for the development of alternate forms of treatment.

Unfortunately, the real situation is unlikely to be thus. The "model" model would assume that we would understand the main features of a disorder, and be able to discover or generate these features in an animal. It would also be assumed that there would be no major differences in biochemical or morphologic features between the animal and man. Since etiologic factors in human psychopathology remain only partially known, models have been based principally on some of the *symptoms* of the disorders. The utilization of analogous symptoms, however, may be misleading, as it has to be assumed that the same behavioral or biochemical mechanisms are responsible for the symptoms in the model disorder and the natural disorder. This assumption may not be valid, and it may well be the exception rather than a common finding that depression, anxiety, schizophrenia, and other psychopathologic states are controlled in different individuals by only one set of variables. As indicated in Table II, many of the current animal models under discussion at this meeting and in the recent literature are incomplete, and

Table II. A Survey of Several Examples of Animal Models for Human Psychopathologic States in Terms of Their Contributions to Understanding Different Components of the Disorders

	Models			
	Separation	Foot shock	Reserpine	Amphetamine
Etiology				
Direct	+		−	?
Vulnerability	(+)		(+)	
Genetic				
Symptomatology	+		+	+
Mediating mechanisms	?		?	?
Treatment response	?	+	±	+

have generally been derived or studied from one point of view, with incomplete information available as to the applicability of the model to questions of etiology, symptomatology, mediating mechanisms, and response to specific treatments.

It may be worthwhile to consider a few examples from our clinical studies, by way of aiming towards some "principles" of the sort Dr. Hinde referred to as a goal of studies of comparative behavior and psychopathology. Table III summarizes some of the issues which need to be kept in mind in discussing animal models for human psychopathologic states. The first point is the most crucial. In order to create an animal model, we must understand some of the major features of the human disorder. Unfortunately, we know relatively little of the etiology of disorders such as depression and mania. We mostly believe that they have mul-

Table III. Issues in the Evaluation of Models for Human Psychopathologic States

I.   Unknown factors in human psychopathologic states:
   A. Uncertain, mixed etiology, with multiple determinants of behavior.
     1. Long-term factors: hereditary vulnerability, early developmental experience, and inter-personal relationship patterns.
     2. Acute factors: life changes, especially separations and losses, conflicts, stresses, and biological changes.
   B. The occurrence of subtypes and large contributions from individual variation in psychopathologic states.
II. Behavioral repertoire differences between man and animals: the problems of reductionistic simplification and anthropomorphic interpretation.
III. Biochemical and anatomical differences between man and animals: problems related to differences in neurotransmitter systems, drug metabolism and behavioral models.

tiple determinants, some of which are summarized here. In addition, we also suspect that there are subtypes of such disorders as depression and mania, and that individual differences may have relatively large influences on the course of these disorders.

For instance, one might have attempted to build an animal model for mania and depression on the basis of symptoms—and might have begun with what commonly are considered to be the cardinal features of these so-called affective disorders, depressed affect or sadness, and elated affect or euphoria. In the case of mania, however, recent phenomenological data, which is summarized below, suggests that euphoria or elated affect is not one of the core features of all manic individuals.

Using a newly developed behavioral rating scale for mania with items which were found in preliminary studies to be highly reliably rated by observers and to reflect global mania ratings done by psychiatrists, a group of items was identified by factor analysis and by their high correlations with global mania ratings that represented what we called "core" manic behavior. Two points were of special note; euphoria and elated affect were not included in the group of items reflecting core manic behavior, but rather only occurred in a subgroup of manic patients. Secondly, most of the manic patients manifested depressed thought content and behavior, including crying and suicidal preoccupations, as part of their manic behavior. In fact, there was an overall positive correlation, which was statistically significant, between severity of mania and a group of items referring to depressive thought content and behavior.

The main point at issue here is that in some ways, studies of these human disorders, including even simple behavioral phenomenology, is still just beginning. It is not surprising that there is some resistance to animal models for these disorders when even the symptomatology, to say nothing of the etiology of these states, is only partly known. There is actually a great need for much more careful study of the simple behavioral phenomenology of these disorders. In fact, it has recently been suggested by a British MRC clinical study unit in Edinburgh that some of the behavioral phenomena observed and rated in animal lesion and drug studies might be evaluated and rated in patients. Phenomena akin to stereotypic activity, exploratory behavior, and aggressive–submissive behavior occur in somewhat altered form in patients, but, in fact, have not yet been subjected to systematic observation and documentation in man. Animal model building may well be outracing knowledge of the phenomena to be observed in natural human states.

Most of our investigational work has been focused on possible etiologic processes in the development of acute episodes of depression or mania, and their reversal by psychoactive drugs. Following leads from studies of the biochemical changes produced in animals by drugs which are effective in these disorders, an apparently coherent pattern of drug effects suggested that depression might be associated with diminished effectiveness of biogenic amines in brain, and mania with enhanced amine function. Our studies have utilized the approaches Dr. Kling has used in his primate studies, namely treating patients with drugs which have relatively specific effects directly on the synthesis of biogenic amines.

Our studies have been most successful in demonstrating patient subgroup differences, and in emphasizing the variety of individual behavioral responses which occur with changes in catecholamine function, rather than in verifying that there is a simple disorder in biogenic amine function in these patients with depression and mania.

To briefly summarize our data from studies of L-dopa, simply increasing brain catecholamines by large oral doses of L-dopa, the amino-acid precursor of the catecholamines, was not found to be associated with striking antidepressant effects. Only 25% of the depressed patients improved, although it was of interest that it was principally one subgroup, patients exhibiting psychomotor retardation, who did improve, whereas none of the patients with the more anxious, agitated form of depressive symptoms improved. However, among the group of patients who did not improve, most were rated as exhibiting more overt anger and irritability during DOPA administration. In addition, those depressed patients who had preexisting evidence of psychotic phenomena (principally depressive and paranoid delusions) as part of their depressive symptomatology were rated as more psychotic during the DOPA treatment period. Of particular interest was the development of hypomanic behavior in the subgroup of patients who had previous episodes of mania—the so-called bipolar patients.

We interpreted this data in the light of other behavioral response data from animals and nondepressed humans as being most compatible with an overall behavioral activating effect of L-dopa. In animals, L-dopa in large doses markedly increases locomotor activity as well as some other more structured behavior such as conditioned avoidance responding. In patients, L-dopa did not seem to have a general antidepressant effect, but rather seemed to produce a similar activated state characterized by an apparent amplification of whatever the preexisting behavioral and psychobiological state was. It is an interesting question to consider whether we are observing here a neurochemically induced adjustment upwards in a set point for psychomotor activity or some similar function in behavioral regulation like that described by Dr. Delgado using electrical stimulation.

I cannot review all the evidence for this conclusion concerning this postulated activating effect of DOPA here, but it seems pertinent to the question of animal models for psychiatric disorders in several ways. First, it does affirm a general role for catecholamines in affect-related behavior—perhaps seen most clearly in the case of the precipitation of typical hypomanic episodes in bipolar patients. Secondly, however, it indicates that the role of catecholamines in depression is complex, and cannot be understood in the terms of one specific animal model for depression—the pharmacologic model based on the reversal by L-dopa of the so-called depressed behavior which occurs in animals and to some extent in humans treated with the drug reserpine. This state can be reversed by many antidepressant drugs, including the tricyclic agents, MAO-inhibitors, and amphetamines. Its reversal by L-dopa had been suggested to establish catecholamine depletion as a possible etiologic model—representing a common metabolic abnormality—in different depressed states. It would now appear that L-dopa's activating effects cannot be equated to the antidepressant effects of other drugs, and that the L-dopa–reserpine interaction probably represents a specific symptomatic model and not an etiologic model for all depressed states. It is very interesting that in some subsequent studies in primates using a different drug, α-methyl tyrosine, to evaluate the catecholamine depletion model for depression, L-dopa again produced only activation and increased aggressive behavior, but not a normalization of impaired social behaviors.

There is not time to review our own clinical studies with α-methyl typrosine and with L-tryptophan, tricyclic antidepressants, MAO-inhibitors, and amphetamines. I would only say that there are many findings in our clinical studies which indicate close parallels in the behavioral and biochemical responses in man to the effects of these drugs in animals.

To conclude, I would like to mention two animal model study approaches that seem especially important to clinical investigators. First, there has been increased interest in the last several years in examining the effects on biochemistry and behavior after longer-term administration of psychoactive drugs. This interest was exemplified in Dr. Irwin's presentation of not only acute but also chronic pretreatment drug interactions with the foot-shock symptom model. This approach is especially important, as many of the most effective drugs used clinically require up to several weeks to become effective, and it now appears clear that extrapolations from acute drug effects in animals to the effects of longer duration treatment in man represent a poor use of the opportunities for model-state comparisons.

Second, another extremely promising trend is the study of the apparently lasting effects of the administration of drugs (such as reserpine and the sex steroids) during early developmental states in animals. It appears from a number of studies that adult behavior or a susceptibility for altered behavior in response to various kinds of stresses can be affected by one-time treatment during critical periods of development. Certainly, if these pharmacologic studies as well as the behavioral studies suggesting long-term effects of early separations on responses to subsequent separations or to other behavioral or biological stresses later in life can be firmly established, a most important principle for the understanding of adult psychopathology may become available.

SUMMARY

1. Animal models of "psychopathology" are vitally needed. From the viewpoint of human psychopharmacology, it is very striking that some of the clinically most effective psychoactive drugs (e.g., lithium carbonate and imipramine) have minimal, nonspecific effects in normal people and animals, but marked effects in patients with affective disorders. Studying normal animals may not help us understand these agents as much as studying them in a valid animal-model state might.

2. An ideal animal model which was a simplified replica of the human disorder in terms of etiology, symptoms and response to treatment is most needed, and would be especially useful for the development of alternate forms of treatment.

3. Major limitations impeding the discovery of useful animal models come from our inexact understanding of the etiology and even of the precise phenomenology of the human disorders. Furthermore, the heterogeneous nature of some psychiatric disorders as exemplified in subgroup differences and other types of individual differences in etiology, symptoms and responses to treatment contribute impediments to the development of generally applicable models, and indicate the need to anticipate that multiple models may be required.

4. Available animal models for depression (e.g., the behavioral separation models, and pharmacologic models based on catecholamine depletion produced by reserpine or α-methyl tyrosine) have some partial validity, but also have major limitations, and only seem relevant to some features of the natural disorder.

5. Recent developments using animals bred for certain behavioral characteristics, animals treated chronically with drugs, and animals treated with drugs, hormones or behavioral manipulations (e.g., separations) at critical periods in early development may provide especially important model situations for understanding such issues as differences in the vulnerability of adults to develop psychopathology in response to psychological and biological stresses.

WORKSHOP III:
Neurophysiological Experimental Modification of the Animal Model as Applied to Man

Edited by DENNIS MURPHY

The afternoon discussants seem to agree that with respect to certain human illnesses, e.g., epilepsy, depression, hyperactivity, or symptoms related to observable motor abnormalities, animal models may be quite pertinent and in some cases, have already been quite useful. However, the development and relevance of animal models of schizophrenia remains controversial and no clear evidence for such a model has yet emerged. Dr. Delgado and others have argued that schizophrenia is a distinctly human illness, manifested by disturbances in thinking, affect, verbal communications, interpersonal relations and, in many cases, hallucinations—none of which can be effectively studied in animals despite our ability to alter their brain chemistry, rearing conditions or social-environmental setting. Perhaps, as suggested by Dr. Delgado and seconded by Dr. Corson and Dr. Serban, animals do not have the same components in the cerebral cortex that might be responsible for the mediation of "schizophrenic" symptoms.

Dr. Arnold Friedhoff mentioned that at present we do not know why mammals have the capacity to make hallucinogens. We believe that the hypothesis

DENNIS MURPHY ● Section on Clinical Neuropharmacology, Laboratory of Clinical Science, National Institute of Mental Health, Bethesda, Maryland. (Workshop moderated by Ronald D. Myers.)

concerning schizophrenia resulting from the aberrant formation of endogenous hallucinogens is somewhat simplistic. Dr. Friedhoff went on to say that it remains an open question as to whether endogenous hallucinogens play any role in schizophrenia.

Investigators working more directly in psychopharmacology seem to be more optimistic and point to symptoms which may be produced in both man and animals by drugs (e.g., D.M.T., amphetamines, phencyclidine). Such symptoms in man include visual hallucinations, disturbances in feeling, activity, perception, affects, and thinking, but are subject to wide individual variations and, perhaps, genetic susceptibility. In some cases, marked behavioral disturbances may be observed with the same drugs in animals (e.g., amphetamines), particularly in motor activity and overt affective states. While no one has argued that in fact an animal model of schizophrenia is available or even on the horizon, certain symptoms occurring in the schizophrenic syndrome can be produced. In this regard, most discussants also agreed that the metabolic pathway for dopamine is worthy of continued intense investigation.

Drs. Pichot, Suomi, and Corson, while not addressing themselves directly to the problem of schizophrenia, suggest that naturally occurring syndromes which resemble human depression as a result of genetic, environmental or interpersonal etiologies may be useful models and can be identified in laboratory subjects, zoos, and by veterinarians in clinical practice. In this regard, Dr. Suomi presented some anecdotes from his laboratory on despair in young monkeys and Dr. Corson on "depression" and "paranoia" in the dog.

Concluding Remarks

BORJE CRONHOLM

Being a clinical psychiatrist, I have not performed any animal experiments. I have, however, been reading about animal experimentation with much interest. Given Hinde's highly skeptical and critical remarks (elsewhere in this volume) as regards drawing conclusions about human behavior from animal experimentation, I feel a little reluctant to elaborate on this interest. All the same, I take the risk and give a few examples.

Two or three decades ago we learned very much from Masserman's studies, illustrating in a quite convincing way the importance of conflict in the pathogenesis of neuroses. There seem to be basic similarities in the ways different mammals react to conflict. Certainly, experiments with animals have been of very great importance for the development of psychosomatic medicine. For instance, I would like to draw your attention to the Russian experiments reported by Lapin and Cherkowich (1971). When a male baboon with the highest rank in a community is isolated in a cage he shows signs of anxiety. They subside, but when he is shown other baboons in the group with a lower rank being fed, or (still worse) another male courting the females, he again displays signs of anxiety and fury. When isolated in this way four to five months, he displays a "neurotic" behavior and also develops somatic diseases: hypertension, coronary insufficiency, and myocardial infarction. Man may be exposed to analogous frustrations. Studies using the methodology of Holmes and Rahe (Theorell, 1970) have shown an accumulation of life changes during the period immediately

BORJE CRONHOLM ● Department of Psychiatry, Karolinska Institutet, Stockholm, Sweden.

preceding myocardial infarction. These findings support each other. I think it's quite useful that we are able to produce stressful situations in animals and to study the results, both as regards changes in behavior and changes in the organism. In my opinion, investigations of the influence of stressful stimuli on behavior and physiology of animals represents a major advance in our understanding of psychosomatic disorders. As regards psychiatric diseases in a restricted sense, the results of animal studies are somewhat more ambiguous, even if the studies reported by Harlow and others are most interesting. But whether the condition seen in the deprived Rhesus monkeys should be called depression or something else is a difficult decision. Perhaps the chimpanzee youngsters who had lost their mothers, described by Jane van Lawick-Goodall (1973), really were depressed in the same sense as man may be.

I was very impressed by Corson's study. His beloved dog—nasty Jackson— was quite interesting from many points of view. I don't know what was wrong with him, perhaps he may be called a psychopath. He was not able to control himself, he was impulsive, did not learn from experience and couldn't stand monotony. If this was due to genetic factors or to a cerebral lesion we don't know. Anyhow, it is interesting that amphetamine had a therapeutic effect as it has on hyperkinetic children. Aged people may also react favorably to centrally stimulating drugs. Their use of caffeine as a sleeping pill is an example. I believe that much could be gained from a study of differential psychopharmacology in animals—and also in man.

As regards studies on brain disorders and brain dysfunction, animal experimentation is very fruitful. I would like to mention an example. As you very well know there are memory disturbances after electroconvulsive therapy, consisting mainly in difficulties in retaining what was learned, both before a treatment and during a short time after. These disturbances can be measured rather accurately both in man and in rats. Of course, you can't use verbal tests in rats, but you can construct nonverbal tests that are analogous, and then you find a striking parallel between disturbances in man and in rats. I think that the more primitive structures, the more primitive and basic functions we are studying, the more useful are animal studies. Such a primitive function as "consolidation of memory traces" is an example (see Cronholm, 1969).

The ethologists do not tire of warning us against extrapolating from animal to human behavior. I think they distrust us a little too much. We are not foolish enough to believe that we *are* rats or apes or monkeys. But we certainly feel that we get fruitful ideas from animal studies that may be used in research on man. I think you should be careful to warn us too much! So, at last, I thank you very much—ethologists and other animal experimenters—for all the stimulating ideas that we clinicians have gotten from you.

REFERENCES

Cronholm, B. Post-ECT amnesias. In Talland & Waugh (Eds.), *The pathology of memory*. New York: Academic Press, 1969, p. 81–89.

Lapin, B. & Cherkovich, G. Environmental change causing the development of neuroses and corticovisceral pathology in monkeys. In L. Levi (Ed.), *Society, stress and disease*. Vol. I, 266–279, London 1971.

van Lawick-Goodall, J. The behavior of chimpanzees in their natural habitat. *American Journal of Psychiatry*, 1973, *130*, 1–12.

Masserman, J. H. Principles of dynamic psychiatry. Philadelphia and London: W. B. Saunders, 1946.

Theorell, T. Psychosocial factors in relation to the onset of myocardial infarction and to some metabolic variables—a pilot study, Dissertation, Stockholm, 1971.

The Significance of Ethology for Psychiatry

G. SERBAN

Theories of human nature attempting to explain man's behavior developed concomitantly with the evolution of our knowledge of the surrounding world. The discovery of physical laws governing the earth permitted the replacement of the medieval divine model of man with the mechanical one. When the new scientific era was propelled by the Copernican–Newtonian cosmological revolution, psychology, as a new science, progressively adopted the mechanical model of man introduced by Descarte, worked out by La Mettrie, Cabanis, and J. Mills and subsequently modified by Freud. Yet, the ambitious plans of Freud to provide a scientific understanding of man fell short of his own expectations. Without diminishing Freud's contribution to psychiatry, we can admit that his project of scientific psychology, which was supposed to reduce psychological processes to quantitatively determined physical laws, ended up in metaphysical explanation (Freud, 1954). The Cartesian laws of *res cogitans* became linked to elusive unconscious motivational drives, thereby changing the mechanical model into a metaphysical one. Conversely, an animal component of human nature was always conceded, although man refused to accept any direct comparison with the animals, in a grandiose vision of himself as a final product of God. In this sense, any mysterious meaning attached to his behavior continued to support his divine origins, which explains why mystical theories are so well entrenched in our thinking.

Yet, Darwin's discoveries in biology led to the study of similarities between some aspects of animal and human instinctive behavior. Finally, the experimental

G. SERBAN ● Department of Psychiatry, NYU Medical Center, New York, New York.

work of Bechterev and Pavlov on conditioned reflexes made behavioral researchers believe that they had at their disposal a new model of man governed by
the same behavioral laws as animals, responing to the same physiological law of
stimulus–response drive theory. It was assumed that the laboratory dog or rat
would be able to explain the whole gamut of human behavior by simple extrapolation and interpretation. From the elementary conditioning work of Pavlov
to the work of Hull (1934) and Skinner (1938) evolved a progressively comprehensive theory of human behavior based on an automatic mechanical operation
of satisfaction of instinctive needs. The Hullian view of the science of man was
to regard the behaving organism as a completely self-maintaining machine. His
mathematical formulation of the basic principles of human behavior were re-
evaluated experimentally under the laboratory conditons by Skinner. Based on
his experiments mainly with white rats, which he found very similar in behavior
to man except for differences in sensory equipment, reactive capacities, and the
field of verbal behavior, he constructed a rigid theory of behavior presumably
measurable within the bounds of natural science and allegedly free of subjective
anthropomorphism. If behaviorism was helpful in broadening our understanding
of the learning aspect of behavior, it failed to give us an integrative concept of
human nature. Human nature can not be reduced to experimental models of
laboratory animals, though some inferential formulation for primitive behavior
can be deduced as such. If, however, any scientific conclusions are to be drawn
from the animal model, the animal should be observed in the natural habitat,
since only under these conditons is it possible to determine its interaction with
the environment and elucidate the nature of the adoptive determinants of its
behavior.

Herein lies the major contribution of ethology to the understanding of some
aspects of instinctual behavior in man. Whereas the experimental animal psychologist tests mainly the animal behavior in terms of organ function, the
ethologist attacks the problem of behavior in terms of its sequences for the survival of the species (Eibl-Eibesfeldt, 1971). The causation and its effect becomes
evolutionally significant as an adaptive mechanism to the environment. In this
context the root and developments of particular patterns of human behavior become comprehensible.

A good example to start with would be the application of the experimental
findings pertaining to the mother–infant interaction of rhesus monkeys (Harlow,
1959) to humans. Their extrapolation has clarified the biological function of
attachment behavior of the infant to his mother. For instance, it was demonstrated by Bowlby (1973) that contrary to the widely held theory, the bond
between mother and infant is not based on the feeding process but on need of
the physical contact of the infant with his mother. Furthermore, disrupted
mother–child ties by separation lead to immediate personality disturbance in the
child manifested mainly by a state of depression (Bowlby, 1958). The attach-

ment behavior persists through the individual's life in the sense that the child's early needs for contact with his mother later becomes translated into the need for proximity to the close ones for protection. In this context, we have learned that attachment behavior is an innate biological need and like other instinctual behavior is associated with individual survival (eating, reproduction, etc.) (Bowlby, 1973). Any disturbance of its maturation could affect long-term personality development of the individual. For instance, one aspect of it quite familiar to psychiatrists is related to phobic reaction in children who sleep alone in separate rooms. Under these circumstances, they are more likely to feel insecure, unprotected and subsequently develop fear of darkness or of being left alone. In fact, in a broad sense, the child's early experience of anxiety as an emotional reaction to separation often extends to any new traumatic situation, gradually leading to neurotic behavioral patterns.

Another contribution brought about by ethological findings closely related to biology of attachment is in the area of human psychosexual development. It is generally accepted that sexual deviations have their origin in childhood and stem from the type of psychosexual experience encountered during that period.

The re-evaluation by ethologists of the old concept of instinct in the light of its environmental influences, namely, that of misdirection of instinctual patterns when activated by inappropriate social releasers at critical periods for that instinct offered us a plausible understanding of some sexual deviations. Transvestism and some forms of homosexuality are particularly elicited by this type of inappropriate social activation of the sexual instinct. In most cases, this happens when the mother encourages feminine conduct and/or dress in the boy. Thus, deviant patterns of behavior then lead to the identification of a boy's sexual needs with those of a girl and as such, may become a main target for homosexual experiences. It is important to note that once the sexual releasing mechanism is activated the pattern becomes established, to the extent to which it is reinforced by man's feminine attitude and previously learned behavior which attracted sexual response from other males. The same does not hold true, however, when this type of psychosexual behavior is socially learned on experimental adolescent basis during the noncritical period of sexual development, and without any background of disturbed gender identity. Under these conditions, the sexual experience is without disturbing psychosexual consequences as exemplified by Kinsey research, which indicated that 40% of the male population experienced, at one time or another, a homosexual act, while only 4–5% have remained homosexual throughout their lives (Kinsey, 1949). Apparently, the orientation of sexual instinct towards homosexuality is controlled by its activation at a critical period of formation of gender identity.

Another aspect of human sexuality on which ethology has cast considerable light through the study of primates is that related to the mechanics of orgasm in both male and female. The sexual act as a reproductive function in animal and

man leads to fixed patterns of sexual behavior specific to male and female, resulting respectively in ejaculation in male and eventual impregnation for female.
In humans, to the extent to which sexuality becomes expressed independently
of reproduction, it has evolved into a pleasurable act of intrinsic meaning. Yet,
whereas the physiological pleasure associated with the sexual act appears to be
assured for men, it may not be of the same significance for women. Why should
these differences exist? Apparently, not all women are anatomophysiologically
prepared to obtain sexual gratification to the point of orgasm through male
penetration. With the advent of contraceptives and minimization of the fear of
pregnancy, a woman seeks to derive the same amount of pleasure from the
sexual act as man. Despite the use of various sexual methods culturally derived
to facilitate orgasm, approximately 50% of women are unable to reach an
orgasm without extensive genital manipulation, apparently because of their
anatomo-physiological construction (Kinsey, 1949). Genital manipulation to the
point of orgasm represents pure eroticism unrelated to male penetration or reproduction. This suggests that women could experience two types of sexual intercourse: one for purpose of reproduction—pleasurable, but not necessarily leading
to orgasm, part of the animal heritage—and another one purely for pleasure, which
might require extensive manipulation without necessary penetration, culturally
learned. The old psychiatric assumption that frigidity in women is mainly due to
psychological factors does not receive support from the clinical and comparative
psychological findings.

Another aspect related to the female's ability to reach orgasm is the duration
of sexual act. In the modern concept of sexuality, man is supposed to sustain
long periods of erection in order to satisfy the woman. Yet this apparently is
contrary to his own nature as demonstrated by Kinsey, who showed that the
average time between penetration and orgasm for man is about 2-3 minutes,
while for a woman the average is 8-10 minutes. We have learned from the animals
that ejaculation follows in a short time (5-30 seconds) after intromission since
the main purpose of the act is that of reproduction (Ford & Beach, 1951). It
would appear that if man is not prepared to change, now, the fixed patterns of
sexuality which he developed evolutionally, just to please the needs of the
modern woman, he faces emotional difficulty. If the existent sexual-
physiological differences are disregarded, the man—woman emotional interaction
would become progressively strained. The man is under pressure to perform
which leads to a sense of inferiority unless he can learn new techniques to sustain long erection while women are frustrated due to unrealistic expectations for
orgasm by penetration regardless of their psychological or anatomophysiological
possibilities.

Not only is the sexual interaction affected by the cultural revolution but also
by the psychosocial interaction between sexes. One of the present cultural con-
tentions is that the woman throughout history has been dominated and exploited

by man. This situation allegedly started with the physical superiority of man and continued because of "a general male conspiracy" to control women for their own economical and social benefit. The evolutionary biological findings do not support this conclusion. The human history quoted by the protagonists of this theory does not support it either (Millet, 1970). For instance, the aftermath of the French and Russian revolutions proved amply that the women gained through political power temporarily, and lost it gradually, by simple lack of interest in political matters. Evolutionary social differences between sexes appear to be present in primates, as well, indicating the original formulation of this pattern. The female and male chimpanzee have different social roles in the organization of the communal life. The females take care of the offspring, while the males, based on a hierarchical dominance order, control the members' interaction, protect the territory and look for better feeding grounds (Goodall, 1968). Primitive human society follows the same patterns while in more advanced ones, the social differentiation between sexes becomes less well defined. This is particularly true in industrial societies, where advances in technology have eliminated some aspects of division of labor. Nevertheless, the basic differential function remains unchanged. Women still are in charge of child rearing while men are involved in the policy and strategy of defending territory and providing food.

When the social roles of male and female become artificially challenged by superimposed unprogrammed cultural factors, the basic structure of marital interaction was affected due to the loss of equilibrium based on specific role identification. Indeed, the psychopathology of modern marriage appears to be produced by the lack of clear identification of sex roles in family interaction. With the newly gained independence of women, the basic emotional cohabitation bond between men and women has undergone changes as well. The protective signals elicited in man by the woman's subdued behavior have been inhibited now by her assertive, competitive attitude. As a consequence, the man feels nonprotective and noncommitted to the woman. This leads to psychological tension which interferes with the partners' ability to interact harmoniously outside of the mutual need of sexual expression.

But perhaps one of the most controversial issues on which the ethologist has thrown light is that of aggression. Culturalists impressed by the differences between societies in the expression of human aggression concluded that aggression is a socially learned behavior. Consequently, they assumed that aggression could be eliminated by a simple educational process (Bandura, 1969). The frustration–aggression hypothesis of Dollard has provided an impetus for a new conceptualization of aggression and overshadowed the significance of aggressive instinct (Dollard, Doob et al., 1939). This theory was considered by many as a substitute for an aggressive instinct. Though the theory has its value in limited situations, when it is applied to hierarchical order and ranks of society as sources of frustration, it loses its meaning if its theoreticians attempt to formulate social policies based on it.

Any elimination of source frustration will require the elimination as well of the producing cause; that is, of the social system itself. Moreover, the whole series of experiments which attempted to validate the theory of frustration–aggression besides their inadequate definition of terms and oversimplification of the problem, failed to shed light on the origin of the aggression.

Conversely, ethologists hav clarified more convincingly the origin of aggression and the possibilities for its social control through discriminative modeling behavior (Eibl-Eibesfeldt, 1971). Ethologists, however, are bringing further evidences by comparative animal behavior that aggression is basically innate. Defense of territorial behavior against the aggression of others is, for instance, a classical example of human behavior which follows similar patterns of primates and other social animals. Delineation of territory is not only an accepted mode of interaction between nations but between individuals as well in their daily social intercourse. Any trespassing of the designated territory leads to acts of war between nations or aggressive behavior between individuals. In the same context, it can be added that our social system based on hierarchy and rank order asserts authority and implicit admission of submission. Interestingly enough, the same mechanisms of controlling aggression could be observed in humans and in animals who developed similar evolutionary patterns for working off the aggression. The ritualization of the conflict by verbal argument and/or appeasement gestures became the most usual form for the symbolic expression of the aggressive feelings between men instead of direct confrontation. Conversely, any build-up of unexpressed aggression could reach violent dimensions, and be acted out in situations which are basically nonprovoking. The lack of opportunity to work off the innate aggressive drives in our daily social interaction will lead to aggressive reactions in unrelated situations which are perceived as threatening to our well-being and used to release the suppressed aggressive stimuli.

Translated into psychiatric terms, this means that a certain amount of aggression is part of daily human interaction and has to be learned to be dealt with. The main handling of it should be based on a symbolic pacification of the aggressive tendencies of the other party. Pacification is reached by various means of social appeasement such as submissive gestures, apologies or crying, just to mention a few. In a more general approach, aggression may be inhibited by social neutralizers, such as love bond and friendship. They are the best known counterforces against aggression and have been fully described by Freud and others. The concept of bonds was fully investigated and supported by ethologists like Lorenz who believes that the ability to form bonds between individuals is universal and is based on a bonding drive "which is our primary motive for socialization" (Lorenz, 1966). However, the bonding drive to control aggression has its own limitations, depending on the shifting allegiance of individuals within the bonding group and the interrivalry between various groups. In addition, two dimensions of the emotional bond, i.e., ranking order and obedience, though operational in animals as inhibitors of aggression raise serious doubts as to their reliability for humans. The primate's social organization is controlled by ranking order, but in humans, this bond mechanism could easily turn to rebellion when used by any one group to gain access to power. The history of social revolutions proved this fact beyond any doubt.

At this point, it may be of interest to mention the experiments of Milgram (1965), who indirectly showed the relative value of social rank and obedience in controlling aggression. He demonstrated that any social group could respond to an innate disposition to submit to authority and to obey order leading to aggressive acts regardless of the consequences for victims. Recent history of wars taught us amply this lesson. In this respect, the cultural learning could play an important role for the individual understanding of the limitation of authority and his obedience to it. Another alternative for control of aggression is that of educational brotherhood recommended by culturalists as a variance of bond drive, and considered by them as a condition sine qua non for survival. Individuals and nations should learn to act on behalf of the species, it is claimed by culturalists; yet, if some nations do not act as such, then the ones who act in the spirit of brotherhood are the ones to be destroyed (Eisenberg, 1972). This spirit of brotherhood was introduced thousands of years ago, yet it did not change man's view of himself.

All these particular aspects of instinctual behavior elucidated by ethologists could be integrated in the more comprehensive concept of "fixed action patterns" genetically determined and specific to the species which explain some general human behavioral responses independent of the environment. The basic conclusion which can be drawn by us is that humans have their own developed fixed patterns of adaptation which, when disturbed, lead to conflictual personality reactions.

This raises another question of how far can human adaptability go by modifying our animal heritage. It is well known that with advancement in technology, man has succeeded in changing too fast, and now his environment has created increasing problems for his adaptation. For instance, the crowding led to the disappearance of individual distance in public places or even in apartment houses; the high level of technology forced large masses of population to live a depersonalized life, manipulating daily for 8 hours a day, neutral, uncommunicating machines which control their lives. Furthermore, the privacy of individual life was invaded by mass media and other means of statistical surveillance, computers, etc., all resulting in sources of conflict and stress. Cultural adaptation becomes the main problem which faces the individual, yet as Eibl-Eibesfeldt clearly indicated, its laws have to follow the same patterns of phylogeny if the individual is to be able to survive. The environment can shape the behavior only within the parameters of limited human behavioral adaptability for which he has been preprogrammed biologically. In the final analysis, the apparent difference in cultural responses in different societies have a basic similarity derived from the common human nature which all men share.

The concept of human nature based on a preprogrammed biological structure came under attack from the extreme behaviorists who believe that human behavior could be formulated and organized by conditioned processes, and from culturalists who believe that human nature is a product of socially learned patterns. Applied to psychiatry, they reached conclusions that "man is his own chief product" and since "there is no solid foundation to the theoretical extrapolation of the instinctivists, ethologists, etc.," man's behavior is a self-fulfilling prophecy of his social image (Eisenberg, 1972). This skeptical view about the contribution of behavioral science to the understanding of man does not refrain them from producing their own theories of the modifiability of the biological set by cultural factors, which unwittingly reinforces the discarded Lysenko biological theories of the late forties. The culturalists' dream of a new man, totally free of aggression and appreciative of social values is based on a two-fold, unrealistic assumption: one, the existence of a social system which does not promote interpersonal conflict; and two, an individual who in case of any conflict will always accept the decision of socially imposed mediators. Interestingly enough, the same theoreticians who are against the concept of biological fixity, specific to the species, are directly or indirectly advocating cultural fixity based on a process of learning, which rewards societal-related values and indirectly perpetuates an imposed social organization. In other words, the whole concept of the new science of technology of behavior advocated by Skinner leads decidedly to increased frustration for the individual trapped in a normative social system of rewards and failures (Skinner, 1971). Psychologically, this utopic society presupposes to function by denying a basic human need, that of true affirmation of himself within a socially free competitive system. If for idealistic consideration allegedly to improve man's chances for survival, we decide arbitrarily what is scientific and what is not about human biological structure, we not only distort our basic knowledge of human nature, but we destroy as well our own aim for species survival, because man will continue to behave according to his nature and not as we would like him to, based on armchair theories. Indeed, the biological and social forces in their interaction are shaping man; what we need to know is how this interaction works phylogenetically and ontogenetically for man's adaptation to environment.

In the context of adaptation, we can mention some interesting attempts made to apply the ethological observations to the understanding of some clinical symptoms of schizophrenia. It is generally known that schizophrenics show a high degree of adaptive deficit to environment. The symptomatology of withdrawal and emotional flatness could be viewed as a developmental inability to make social bonds and to have meaningful social relationships (Staehlin, 1953).

Indeed, clinical research does indicate that schizophrenics have problems in adjusting to environment, difficulty in coping with the familial and social stress since preadolescent years (Serban & Woloshin, 1974). If a genetic predisposition is considered to play a significant role in the etiology of schizophrenia, it means then that the normal developmental adaptive mechanism of schizophrenics is disturbed by his innate inability to learn the coping mechanisms in response to social stress (Serban, 1975).

Perhaps herein lies the greatest help offered to us as behavioral scientists by ethologists . . . "a more comprehensive and scientific method of studying human behavior." As Tinbergen stresses, "Behavior is a life process, its study ought to be part of the mainstream of biological research" (Tinbergen, 1968). It is to the merit of ethologists to introduce the biological method to the study of the behavior itself, giving a new perspective for approaching methodologically any research on man. Tinbergen's four principles of ethology are worthwhile for psychiatry as well. They can be extrapolated as follows. What is the significance of a particular behavior for the survival of man? What are the reasons for the presence of a behavior at a given time? What is its psychophysiological mechanism? What is the ontogenesis and phylogenesis of any behavioral pattern? The application of this methodological inquiry into human behavior will permit us a more objective appraisal of man free of any extremist interpretation, be it at the metaphysical or mechanical end of the behavioral spectrum.

In conclusion, for psychiatrists, innate behavioral characteristics should be studied within the framework of cultural influences to the extent to which the process of learning is selective and predetermined by the genetic endowment of the individual. Yet, it would be an oversimplification to extrapolate literally the findings of the animal model to the complex human reality. As Hinde concludes, it would be a simplistic belief to assume that "the capacities of the higher organisms could be understood directly in terms of elementary functions." (Hinde, 1975). The similarity between particular behavioral aspects of animals and humans does not necessarily presuppose that the obtained animal findings could sufficiently and adequately explain man's behavior. The reason is obvious— the animal model of behavior does not take into account the subjective experiential world of behavior. There is a tendency among ethologists similar to that of behaviorists to equate the cause of behavior with its physiological substratum. Yet, subjective factors, as we know, enter into the causation of behavior in humans and to some extent in animals as well. A typical example in this direction in man is provided by the conversion reaction which is a physiological and behavioral response to an emotional subjective–autosuggestive phenomenon. Since the cardinal problem of psychology is that of the interrelationship between mind and body, the simultaneous consideration of the subjective and objective aspects of the same phenomenon becomes a condition sine qua non for the understanding of a particular behavior.

This raises the final question, whether it is possible to construct any relevant and comprehensive model for human behavior, taking into account our present knowledge in all the interrelated biosocial and biochemical sciences. From what we know, it is more likely to assume that we can derive only general laws of human conduct. Their predictability for behavior is limited when applied to groups and even more unreliable when intended for prediction of individual behavior in new and unexpected situations. If we can draw conclusions about "man's behavior adjustability outpaced by culturally determined changes," (Tinbergen, 1968) we are unable to gain any insight into why people act differently under the same circumstances without recourse to the study of the individual himself. The human factors, individually stylized, affect one's behavior as perceptualized by an individual in his own conceptualization of his conflictual environmental world to which he has to adjust developmentally within his genetic heritage. In this sense, man cannot be reduced to any behavioristic or biological formula because due to his intelligence and creativity, he always reaches new, amazing, and unpredictable solutions.

REFERENCES

Bandura, H. *Principles of Behavior Modification.* New York: Holt, Rinehart and Winston, 1969.

Bowlby, J. The nature of the child's tie to his mother. *Int. Journal of Psychoanalysis,* 1958, *39,* 350–373.

Bowlby, J. *Attachment and Loss.* Vols. I and II. New York: Basic Books, 1973.

Dollard, J. C. Doob, L., Miller, N., Mowrer, O., & Sears, R. *Frustration and Aggression.* New Haven: Yale University Press, 1939.

Eibl-Eibesfeldt, I. *Love and Hate.* New York: Holt, Rinehart and Winston 1971, pp. 1–40.

Eisenberg, L. The human nature of human nature, *Science,* April, 1971.

Ford, S. C., & Beach, A. F. *Patterns of Sexual Behavior.* New York: Ace Books, 1951, pp. 42–46.

Freud, S. *The Origins of Psychoanalysis: Sigmund Freud's Letters.* New York: Basic Books, 1954, p. 355.

Goodal (van Lawick), J. The behavior of freeliving chimpanzees in the Gombe Stream Reserve. *Animal Behavior Monograph,* 1968, 1 part III.

Hinde, R. A. *The use of differences and similarities in comparative psychopathology.* In press.

Harlow, H. The nature of love. *American Psychologist, 13,* 673–685.

Hull, C. L. *Principles of Behavior: An introduction to behavior therapy.* New York: Appleton, Century, 1934, pp. 24–26.

Kinsey, A. C., Pomeroy, N. B., & Martin, C. *Sexual Behavior in the Human Male.* Philadelphia: W. B. Saunders, 1949.

Lorenz, K. *On Aggression.* London, Methuen, 1966.

Milgram, S. Some conditions of obedience and disobedience to authority. *Human Relativity,* 1965, *18,* 59–75.

Millet, K. *Sexual Politics.* New York: Doubleday, 1970, pp. 220–233.

Serban, G. Parental stress in the development of schizophrenic offspring, *Comprehensive Psychiatry,* February, 1975.

Serban, G. & Woloshin, G. Relationship between pre- and post-morbid psychological stress in schizophrenics, *Psychological Reports,* 1974, *35*, 567–577.

Skinner, B. F. *The Behavior of Organisms: An experimental analysis.* New York: Appleton, Century, 1938, pp. 45–50.

Skinner, B. F. *Beyond Freedom and Dignity.* New York: Bantam Books, 1971.

Staehlin, Von B. Regulation in the social life of severely mentally ill, *Schweiz. Arch., Neurol. and Psychiatry,* 1953, *72*, 277–298.

Tinbergen, N. On war and peace in animals and man, *Science,* July 28, 1968.

Index of Names

Ainsworth, M.D.S. 28, 34, 35, 62, 63, 65
Anderson, O.D. 117
Axelrod, J. 240

Bechterev, V.M. 278
Bell, S.M. 40
Bender, L. 39
Blehar, M. 39
Blurton-Jones, N.G. 44
Bourne, P.G. 179
Bovet, D. 254
Bowlby, J. 10, 20, 39, 41, 53, 278
Bucy, P.C. 262

Cannon, W. 116
Charlesworth, W.R. 81
Cherkowich, G. 264
Collins, M.L. 16, 21
Corson, S. 68, 175, 177, 181, 265, 273, 274

Darwin, C. 277
Davies, L. 25
Delgado, J.M.R. 72, 269, 273, 369
Denenberg, V.H. 191
Denhoff, E. 127
Descartes, R. 53, 277
Dollard, J. 281
Domek, C.J. 12
Domino, E.F. 265

Eibl-Eibesfeldt, I. 63, 65, 175, 179, 182, 278, 282, 283
Engel, J. 129
Erikson, E.H. 85

Friedhoff, A. 273
Fuller, J.L. 116

Gatti, G.L. 254
Gerard, R.W. 111
Goodall, J. van-Lawick 264, 281

Hamburg, D. 177
Hanson, E.W. 10
Harlow, H.F. 15, 72, 73
Heinicke, C. 28
Himwich, H.E. 247
Hinde, R.A. 12, 15, 21, 29, 34, 39, 63, 64, 66, 72, 182, 266, 267, 285
Holmes, T.H. 264
Holmsledt, B. 233
Hull, C.L. 277, 278
Hunt, H.F. 183

Irwin, S. 66, 67, 182, 165, 270

James, W.T. 117

Kallmann, F. 71
Karczmar, A.G.
Kaufman, I.C. 12, 21, 25, 37, 194
Kinsey, A.C. 280
Kling, A. 67, 69, 183, 269
Kluever, H. 262
Kolb, L.C. 71, 184
Konner, M.J. 42, 44, 46
Kuehl, F.A. 241

Lapin, B. 264
Laufer, M.W. 127
Lehrman, D.S. 81, 190, 194, 195
Levine, S. 63, 65, 66, 73, 179, 183
Lindgren, J.B. 240
Lorenz, K. 28, 30, 41, 46, 80, 81, 94, 96, 282
Lutkins, S.G. 177

McKinney, W.T. 20, 21
Maier, S.F. 142, 156, 166, 167
Main, M.B. 45, 46, 67
Mandel, L.R. 241
Mason, J.W. 179
Masserman, J. 73, 176, 183, 264
Meltzer, H. 183
Meyer, D.R. 129
Milgram, S. 283
Miller, N.E. 165
Money, J. 64, 65, 71, 72
Montagu, M.F. 95

Overmier, J.B. 142, 152, 166

Pavlov, I.P. 2, 227, 278
Pichot, P. 5, 274

Rahe, R.H. 264
Rees, W. 177
Richardson, D. 142
Robertson, J. & J. 28, 38
Rose, R.M. 179
Rosenblatt, D.S. 189
Rosenblum, L.A. 12, 21, 37, 194
Rosenzweig, S. 54, 59, 241

Sachar, E.J. 178
Scott, J.P. 116
Scudder, C.L. 142
Seay, B. 10
Seligman, M.E.P. 142, 143, 149, 151, 152,
    166, 167, 168, 170, 172
Serban, G. 273
Shein, H. 183
Sherrington, Sir Charles 3
Sidowski, J. 61, 69, 70
Skinner, B.F. 81
Spitz, R.A. 10, 28, 41
Stockard, C.R. 117
Stone, E.A. 151, 152
Stroebel, C. 69, 70, 179, 183
Suomi, S.J. 37, 38, 40, 41, 61, 70, 183,
    274

Tinbergen, N. 2, 28, 45, 67, 79, 285

von Holst, E. 94

Weiss, J. 176, 179, 183
Wender, P.H. 117
West, L.J. 183
Wolfe, K.M. 41

Subject Index

Abandonment, 34, 79
Ablation penis, 56
Adam principle, 56
Adaptation
    normal, 101
Aggression, 20, 104, 187, 197, 208, 281,
        282
    biological, 93, 182
    in dogs, 90
    interspecific, 90, 92
    intraspecific, 90
    fighting in,
        Eskimos, 92
        Kalahari, 92
        Iko Bushman, 92
        rattle marine iguanas, 90
        rodents, 90
        snakes, poisonous, 90
        Waika Indians, 92
Anaclitic depression syndrome (*see*
        Depression)
Androgen, 56
Anxiety, 35, 39
    hypermotility anxiety, 209
    pathological anxiety, 33
Appeasement (*see* Phylogenetic adaptation)
Approach–withdrawal play prior to
        maternal separation (*see* Separation)
Arrest reaction, 208
Attachment behavior, 14, 279
    attachment–separation behavior, 50
    phobic reaction, 279
Autistic children (*see* children), 66, 67
Avoidant escape deficit, 143, 146, 149,
        151, 153, 154
Avoidant toddlers, 45

Battered-child syndrome, 49, 184
Behavior modification, 53
Behaviorism, 277, 278
Bereavement, 41, 178
Biodynamic principles of behavior, 101
Bond attachment in animal and human
        models
    attachment behavior, 14
    disturbance of
        chamber confinement, 15
        diphasic-protest despair reactions,
            13
        physical separation, 15
        protest despair, 15
        self clasping and huddling behaviors,
            15, 20
        severe maturational arrest, 14
        social isolation, 22
Bonding, 88
    drive for, 272
    peer pair, 58
Brain
    research in animals, 204
    stimulation in humans
        arrest of the hippocampus, 212
        caudate nucleus, 213
        cerebral disturbances in man, 209
        chemitrode system, 214
        dialytrode bag, 214
        epilepsy, 209
        glia capsule, 211
        hypermotility anxiety, 209
        results of chemical studies, 214
        stimoceivers in man, 211
        techniques for chemical studies, 213
        transdermal dialytrode, 215

293

Brain *(cont'd)*
  electrical stimulation of the, 204, 207,
      209, 210
    chemitrode system, 214
    dialytrode bag, 214
    stimoceiver, 205
    stimoceivers in man, 211
    techniques for electrical studies, 204
    telemetry of electrical signals, 205
    telencephalic region of the, 162
    transdermal stimulation of the human
        brain, 204, 206
  electroencephalography, 205
    in cats, 207
  microminiaturized batteryless stimulator,
      206
  techniques for electrical studies, 204
  telemetry of electrical signals, 205
  theta rhythm, 207
  ventricular infusion in brain of mice,
      146, 147

Central Kalihari Bushman, 51, 63
Children
  fear in, 84
  hyperactivity in, 67
  hyperkinesis *(see* Hyperkinesis, in
      children)
Chimpanzee *(see* Primates), 78
Chromosomally male intersexes, 51
Cultural adaptations
  bonding, 88
    and cultural patterns, 95
    and cultural relativism, 96
  pseudospeciation in man, 93, 94
  signals of children, 86
  specific patterns
  subspeciation in man, 93

Depression, 11, 22, 28, 69, 264
  anaclitic depression syndrome, 10, 17,
      18, 20, 21
  animal models for mania and, 268, 270,
      271
  depression or mania and their reversal by
      psychoactive drugs *(see* Psychoac-
      tive drugs)
  depressive self directed behavior, 14, 15
  in mother—child separation *(see* Separation)
  overtly depressive behavior, 21
  reactive depression in humans, 170, 171

Depression *(cont'd)*
  in rhesus monkeys *(see* Depression,
      animal models)
Deprivation, 40, 59
  early, in animals, 102
Dopamine, 170, 171
  dopaminergic control, 18, 234
  release, 225
Drug effects
  drug response to stress
    amphetamines, 67, 152
    avoidance—escape test, 153, 155, 169
    brain neurotransmitter activity, 151,
        176
    infusion of norepinephrine, 152
    noradrenergic activity, 149
    noradrenergic deficiency, 144
    norepinephrine, 146, 149
    radioactive ^{3}H-norepinephrine, 146,
        147, 148
    tricyclic antidepressants, 152
    tryptophane hydroxylase, 221
  in hyperactive children
    amphetamine, 67
    anticholinergics, 18, 232
    antihistaminics, 232
    catecholamine releasing major tran-
        quillizers (reserpine and tetra-
        benazine), 231, 232
    D-amphetamine, 67
    dopaminergic control, 18, 234
    L-dopa (increasing brain cathechola-
        mines) 269, 270, 273
    magnesium pemoline, 18, 232
    major tranquillizers, 18, 231, 233
    minor tranquillizers, 18, 232
    ethyl alcohol, 18, 233
  narcotic analgesics, 18, 232
    narcotic antagonists, 18, 231, 234
    O-adrenergic blockers, 18, 232
    O-methyl typrosine, 270
    pharmacologic agents producing model
        psychopathologic syndromes, 266
    reserpine, 270
    sedative hypnotics, 10, 232
      in relation to foot-shock stimulation,
          233
    therapeutic correlates of drug action,
        235
    tricyclic antidepressants, lithium, 10,
        18, 67, 231, 265, 266

Drug effects *(cont'd)*
  on induced agitation in mice
    amitriptyline, 228
    apomorphine, 221, 222, 223, 236
    benactyzine, 230
    chlorpromazine, 225
    chloridiazepoxide, 220, 223, 225, 227
    chlorophenylalamine, PCPA, 221, 222
    cyclazocine, 227, 228
    d-amphetamine, 225
    imipramine, 222
    meperidine, 227, 228
    meprobamate, 230
    methadone, 227, 228
    morphine, 227, 228
    PA postural arrest, 221, 222, 226, 227,
      228, 229, 230, 232, 233, 234, 236
    perphenazine, 223, 225, 226, 228
    reserpine, 222
    serotonergic or dopaminergic
      mechanisms in behavior, 221
    SMA spontaneous locomotor activity,
      221, 222, 227, 228
Dwarfism
  hyposomatotropism, 49
  psychosocial dwarfism, 49

Early psychological trauma, 102
Environment
  as a constant in experimentation, 192
  nuclear family social, 18
  post-separation *(see* Separation)
  pre-separation *(see* Separation)
  reunion, 15
Emotions
  emotional interaction between man and
    woman, 280
Ethology, 278, 285
Experimental neurosis and psychosis in
    animals, 105

Frustration—aggression hypothesis, 281

Gender-identity differentiation, 50, 56
Genetic hierarchical structure, 72

Hermaphrodites, 51

Homosexuality, 279
  in monkeys, 71
  in twins, 71
Human psychosexual development, 279
Hyperkinesis
  in animal models
    conditioning experiments on dogs,
      118, 120, 124, 131
    diuretic and antidiuretic dogs, 116, 117,
      124
    dogs with experimentally induced
      anxiety, 112
    flight or fight reaction, 116
    hyperkinesia, 113, 119, 122; in dogs,
      69, 181
    MBD (Minimal brain dysfunction) syn-
      drome, 114
    platelet serotonin levels, 114
    urinary catecholamines and electro-
      lytes, 118
  in children
    amphetamine, 67
    catecholamine transmitter deficiency,
      132
    childhood hyperkinesis syndrome, 117,
      135
    hyperactivity, 67
    hypoinhibitory syndrome, 131, 132
    minimal brain dysfunction in children
      abnormal children, 113, 131, 133
      EEG, 130
    specific learning disability (SLD), 133
    drugs for *(see also* Drug effects, in
      hyperactive children)
    chlorpromazine, meprobamate, am-
      phetamine, 111, 118, 120, 121,
      124
    CNS stimulants (amphetamine and
      methylphenidate), 114, 124
    d-amphetamine, *d*-isomer, *l*-ampheta-
      mine, 127, 128, 129, 130, 132
    Dilantin (diphenylhydantin), 124
    haloperidol, 130
    meprobamate used on anxiety, 112
    monamines, 114
    norepinephrine, 129
Hypomanic behavior
  in bipolar patients, 269
Hyposomatotropic dwarfism *(see* Dwar-
    fism), 49
Hypospadiac phallus, 51

Idiodynamic *vs.* phylodynamic, 54, 55
Induced foot shock fighting
    in mice, 219, 225, 234

Kissing
    in man, 79
    in monkeys, 78
Klinefelter's syndrome, 55

"Learned helplessness hypothesis," 143,
    151, 155, 157, 158, 160, 168

Marriage
    psychopathology of modern marriage, 281
Maturational arrest, 21
Mice (*see* Stress coping behavior in mice;
        also Drug effects, induced agita-
        tion in mice)
Microthermistors, 164
Monkeys (*see* Primates)
Monoamines, 149, 165
    MAO oxidase, 166, 179
    pargyline (monoamine oxidase inhibitor
        165, 167)
Mother–infant
    interaction, 12
    relationship, 194
Motivation, 101
Motor activation deficit theory hypothesis,
        144, 153, 154, 158, 160, 165,
        168, 170
    cold swim, 150, 154
    hypothermia, 151, 152

Neurotigenesis, 101
Norepinephrine, 148, 157, 158, 160, 162,
        167, 170
^{3}H-norepinephrine, 162, 170

Paraphilia, 50, 58
Pavlovian school, 117, 118, 120, 125
Peers
    directed activity, 13, 14, 15
    multiple short term separation of peers
        (*see* Separation)
    pair bonding (*see* Bonding)
    play, 10
Perception, 101
Photo-metrozol test, 113

Phylogenetic adaptation
    appeasement gestures (verbal agreement
        ritualization of conflict), 282
    appeasement in cormorants, 78
    chimpanzees, 78
    conspecifics, 84, 182
    cormorants, 77
    environmentalism, 83
    ethology, 77
    homology, 78
    imprinting, 83
    insects, 79
    instinct, 81, 82
    interactionists, 81
    internal motivating mechanisms, 83
    Japanese macaques (monkeys) 84
    learning in animals, 83
    Masai greeting, 80
    motor patterns, 82, 182
    ontogeny, 81
    pseudospeciation, 85
    releasing mechanisms, 82
    signal, 85
    social releases, 82
Primates
    (*see also* Bond attachment in animal and
        human models)
    living with peers, 13
    primate models,
        bonnet, 9
        macaque, 41, 72, 196, 198
        pigtail, 9
        rhesus, 9
    ranking order, 283
    rehabilitation of monkeys, 283
    social differences between sexes, 281
    social separation in monkey infants, 12
    syndromes of psychopathology, 24
Psychoanalytic doctrine, 53
Psychosomatic disturbances, 105

Radio frequency transmitter (RF), 204
Reserpinized rat, 9

Schizophrenia, 284, 285, 273
    animal model – indole hallucinogens
        BFT bufotenin, 5-hydroxydimethyl-
        tryptamine, 239

Schizophrenia *(cont'd)*
  animal model *(cont'd)*
    DMT dimethyltryptamine, 239, 241
      243, 245, 249
    hallucinogenic indoles tryptamine, 239
    indole hallucinogens and schizophrenia,
      239
    MAO inhibitors, 240
      L-methionine, L-tryptophan TP, 240,
        245, 249, 253
    MTHF 5-methyltetrahydrofolic acid,
      241
    NMET *N*-methyltryptamine, 241
    NMT, *N*-methyltransferase, 240
    perphenazine, 230
    SAH *S*-adenosylhomocysteine, 242
    SAM *S*-adenosylmethionine, 241
  patients, 243, 244, 245, 252
  preventive formation of hallucinogenic
    indoles
    chlorpromazine, 252
    chromotographic assays, 243
    d-amphetamine, 252
    DMT, (dimethyltryptamine), 246, 250,
      254
    hyperthermia, 253
    LSD (lysergic acid diethylamide), 244,
      245, 252, 253
    MAO enzymes, 249, 257, monamine
      oxidase
    methionine, 249
    methiothepin effective antagonists of
      DMT-induced hyperthermia, 253
    Phenylalanine, 247
    preventive formation of hallucinogenic
      indoles, 246
    SAH *S*-adenosylmethionine, 249
    schizophrenic patients, 243, 244, 245,
      252

Schizophrenia *(cont'd)*
  preventive formation *(cont'd)*
    TP die-trypthophan, 247
    decarboxylation to T, 248
    tyrosine levels, 247
      AAAD aromatic amino acid decar-
        boxylase, 248
Separation, 10, 13, 18, 20, 23, 38, 40, 70,
  178
  anxiety, 62
  between mother and infant, 194
  environment, 15, 23
  fear of, 33
  multiple short-term separation of peers,
    14
  pre-separation environment, 22
  protest despair reaction to, 15, 20
Sex research, 279
Sleep
  EEC patterns during, 70
  rapid eye movement (REM), 71
Stereotaxic surgery, 204
Steroids, 163, 178, 179
Stress coping behavior in mice
  adrenal steroids, 148
  "avoidance—escape deficit," 143, 146,
    149, 151, 153, 154, 165
  neurochemical habituation, 149
  stressors, 149
  ventricular infusion (in brain of mice),
    146, 147
Stressors test, 150, 159, 160

Transvestism, 279
Turners syndrome, 55
Tyrosine hydroxylase activity, 158, 161,
  162

Weaning, 196